T0209966

First

Snow

on

Fuji

First Snow on Fuji

Yasunari Kawabata

TRANSLATED BY MICHAEL EMMERICH

COUNTERPOINT
BERKELEY

Library of Congress Cataloging-in-Publication Data

Kawabata, Yasunari, 1988–1972.
[Fuji no hatsuyaki. English]
First snow on Fuji / by Yasunari Kawabata ; translated by
Michael Emmerich.

    p.   cm.

  ISBN 978-1-58243-105-5
  I. Emmerich, Michael. II. Title.
PL832.A9F813   1999
895.6'344—dc21

                                                          9935103
                                                              CIP

Printed in the United States of America

COUNTERPOINT
2560 Ninth Street
Suite 318
Berkeley, CA 94710

www.counterpointpress.com

Contents

Translator's Note

Kawabata's fifty-ninth birthday was only a month away when *First Snow on Fuji* was published in April 1958. He'd already written most of the novels that would come to be known as his late masterpieces, and though he wasn't yet as widely celebrated in the West as he was in Japan, it must have been clear that his reputation as an international writer would only continue to grow as translations of the quietly stunning later works appeared. It must also have been clear that as his fame spread, and as he aged, Kawabata would be forced to slow down, to write less. His time would be less and less his own.

And indeed the foreseeable occurred. Kawabata was hospitalized for more than six months after suffering a severe gallstone attack in November 1958. He spent much of 1959 and 1960 traveling through Europe, accepting awards and attending conferences. And in 1960—for the first time in all his years as a writer—he published nothing. He was hospitalized for a second time in 1962 when efforts to end an eight-year-long addiction to sleeping pills landed him in a coma. Ten days passed before he woke. But then he was out in the world again, striving to make Japanese literature international, doing everything he could to help other writers. He himself had extremely little time to write. As Kawabata the writer became increasingly visible, Kawabata the man who wrote did indeed vanish into a certain kind of silence.

If this trend was clear to anyone in 1958, it must have been obvious to Kawabata. He'd been the busy president of Japan's P.E.N. Club for a decade. He'd spent two years organizing P.E.N.'s twenty-ninth international conference, which was held in Tokyo in 1957. He could already see what was coming. Kawabata wrote a brief essay on the final day of the 1957 conference he'd worked so terribly hard to prepare. "The 'I' in me seems to have disappeared," he writes. "Or perhaps I ought to say that a different 'I' has been living inside me."

This is when *First Snow on Fuji* was written. A time when Kawabata—still at the very height of his powers—could count the minutes he called his own. He had to make the most of each unclaimed moment, each precious word. So it's no surprise to find that the pieces in this collection are incredibly distilled, often dealing with the relationship between language and being, words and the past, and with being claimed, with losing possession of one's historical self. The book was published at a pivotal moment, and we find traces in these stories of a great many of Kawabata's earlier writings. We find its author looking back, responding to his past, even as he faces his future.

Readers familiar with Kawabata's "Diary of My Sixteenth Year" may feel as if they've come across the third story in this collection, "Nature," before. Kawabata had lost both his parents by the time he was three, his grandmother and his younger sister by the time he was ten. He lived alone with his blind, eventually bedridden grandfather until the old man passed away, and the life Kawabata lived as he waited for that death—feeding his grandfather, helping him urinate, running from the house filled with anger and running back full of sadness—this life is the principal subject of "Diary of My Sixteenth Year." Perhaps "Nature" could be read as a much older Kawabata's postwar revision of that extremely early work, a fresh attempt to lay an old ghost to rest. Only the story's structure is more complex, the dying man is now someone else's grandfather, and the narrator knows nothing but what he's been told.

Kawabata wrote in his 1950 afterword to "Diary of My Sixteenth Year" that what most intrigued him as he looked back over the piece was "the mystery of my being unable to recall these things, even though I once experienced them." This mystery is deeply embedded in "Nature."

One Japanese critic suggests that Kawabata wrote "Yumiura" as a sequel to another very early work, "The Izu Dancer." He believes that the woman who shows up at the aging author's door in "Yumiura" is the earlier story's childish dancer, now grown into a woman. "The Izu Dancer" is thought to be a fiction spun from real events; perhaps "Yumiura" has a sort of camouflaged truth in it too? Is it a particularly personal work? Some readers will be hesitant to accept such speculations, but of course speculation has its pleasures. Kawabata included both "The Izu Dancer" and "Yumiura" (along with "Chrysanthemum in the Rock") in a collection of his own favorites from among his works that was published not long before he received the Nobel Prize. The book includes so few short stories that it seems reasonable to assume—this if nothing else—that Kawabata was exceptionally fond of the stories, that "Yumiura" and "The Izu Dancer" were indeed linked in one way or another within the realm of their author's affections, perhaps even in the manner suggested.

Much more could be said about the stories in this book and their ties to Kawabata's other writings, to his biography, and about the lines that string the stories together, since their sequence is not chronological. It is my hope that this first translation will prompt readers and critics to begin that saying. As for myself, I'm unable to conclude without at least mentioning the story that first took hold of me and made it necessary for me to translate *First Snow on Fuji*. My first reading of "Silence" left me dumbstruck. Such an intense expression of Kawabata's aesthetics and thinking . . . of the "nothing" he discusses . . . of so much that lies at the heart of postwar masterpieces like *The Master of Go* . . . of the meaning of the fallen chestnut that Shingo chooses not to mention to his wife in *The Sound of the Mountain* and the everlasting presence in their marriage of the "blank" his silence leaves. "Silence" was a central work in Kawabata's oeuvre. As I read it again it occurred to me that it was also eerily prophetic. In the story a young writer takes a cab from Kamakura, where he lives, to the neighboring town of Zushi. There he visits another writer, an older man who has stopped speaking and writing, who has given himself over to silence. Kawabata lived in Kamakura and had a studio in Zushi where he went to write. It was in the Zushi studio that he killed himself in the spring of 1972. He left no note. An obituary quoted him as having said, "A silent death is an endless word."

Kawabata probably had no intention of dying when he wrote "Silence" in 1954, though it does seem likely that he foresaw the coming of a different sort of silence, the silence of celebrity and the silence of sickness, the silence of the writer who spends his time making it possible for others to write. He foresaw its coming because his own generosity was finally responsible, because he himself took up the burden of silence. *First Snow on Fuji* arises from silence.

One final note. "The Boat-Women: A Dance-Drama" is one of only two works for the stage that Kawabata created. The "verses" it contains are neither "poems" nor "songs"—they are instead some combination of the two. Each verse is marked with a vaguely mountain-like symbol, which indicates that the line is to be intoned (or maybe chanted?) rather than simply spoken. The verses are divided into blocks of five or seven syllables in the Japanese, and they are written in an older form of the language; the rest of the dance-drama is composed in fairly contemporary prose. The battle fought between the enemy Genji and Heike clans at Dan-no-Ura in 1185, the history of which is told in *The Tale of the Heike*, forms the background of the piece. It is perhaps helpful to know that at this battle the Genji triumphed over the more sophisticated and courtly Heike, and that the story of Dan-no-Ura is one of the most moving and often-told stories in Japanese literature.

So many people have helped make this book possible that I can hardly begin to thank them all here.

Richard H. Okada, Joyce Carol Oates, and Karen Emmerich made any number of valuable suggestions during the early stages of translation; Peter Constantine and Jennifer Lyons provided much-needed moral support; Bradford Morrow created opportunities. My Japanese friends and teachers, especially those at the Kyoto Center for Japanese Studies, created potential. My parents made every potential possible.

This
Country,
That
Country

1.

Takako read "This Country, That Country" in the *Sankei Daily Times* a second and even a third time on the eve of Culture Day, which is to say on November second. The column printed curious and interesting articles about occurrences abroad, more like stories or seeds of stories than hard news.

The previous day's edition had given rather extensive coverage to an announcement made by England's Princess Margaret, in which she had said that she would not marry Group Captain Townshend after all. It was only natural that one of the stories in today's "This Country, That Country" should concern the princess's love affair:

One often comes across mounds of stones in the Scottish highlands. In the past, these mounds were erected in memory of heroes who fell in battle, but now it's said that lovers who add stones to these mounds achieve "eternal love." Four years ago, at a time when Princess Margaret and Group Captain Townshend were both staying in Balmoral, they placed a stone on a mound located in the middle of an overgrown field some three miles outside of town, swore their love for one another, and by this act leapt instantly into fame. The princess's love affair has now ended.

There was a picture of the mound at the end of the article. Its size could be estimated from the size of the people who stood around it—the pile itself was almost as tall as a man, and the individual stones that formed it were a good deal larger than a person's head. A few stones were as wide across as a person's shoulders.

Of course it was impossible to tell which of the stones the princess and the group captain had placed on the pile, but none looked as though the princess could have lifted it alone. She and Group Captain Townshend must have lifted the stone together, and even so it must have been heavy.

Gazing at the photograph, Takako tried to imagine the princess as she would have looked hoisting a stone onto the mound with the group captain, but the image that came was simply an image. Takako felt no connection to it. Her reading of the articles in the previous day's paper had left her feeling sorry for the princess, who had after all been forced by church law and by certain customs of the English royal family to abandon her love, but that feeling was now gone. In some ways yesterday's empathy itself seemed like a foreign story.

Takako was unable to read one of the other stories in "This Country, That Country" with so much detachment, however. The story described two actual cases of "spouse swapping."

The first incident had occurred in Sweden. Two married couples, the Polsens and the Petersons, lived in adjacent apartments in a single building in Egresund, a town near Stockholm. Mr. Polsen and Mr. Peterson were friends of long standing, and they and their wives had grown so close that they lived essentially as a single family. Then, on the twenty-ninth (the article ran on November second, so it must have been October twenty-ninth), the two husbands swapped wives—or to look at it from the other point of view, the wives swapped husbands. In short, the couples were divorced and remarried simultane-ously. Neither the Petersons nor the Polsens are at all worried about the shock they've given the world, and all four of them are getting along as well as ever, it was reported.

"There are so many marriages that just aren't happy, where the couple would be better off getting divorced," Peterson stated. "There's really nothing strange about our marrying each other's wives. In the end it seemed it'd be better for the children—that's basically why we did it."

The Petersons have one child, the Polsens two infants. All three children accompanied their respec-tive mothers when they moved, each into the apart-ment next door.

Another swap took place in the United States in the state of Wisconsin, where on the twenty-eighth (probably the twenty-eighth of October) a spouse-swapping ceremony was held. Forty-three-year-old Mr. Pierce and twenty-nine-year-old Mrs. Pierce married thirty-two-year-old Mrs. Pemis and thirty-two-year-old Mr. Pemis, respectively. The two wed-dings were held consecutively, and each couple helped out at the other's ceremony.

The couples were interviewed two days later, on the thirtieth.

"All of us, our children too—we're all extremely pleased."

Each family has three children. As in Sweden, the children followed their mothers, who were the ones to move, though here too it was hardly a "move" at all: the couples live in facing houses, on opposite sides of the same street.

The comedy of the (very likely) middle-class spouse-swappers affected Takako more powerfully than the tragedy of the princess and her mound of stones because it related directly to her own life.

Or could it be that the spouse-swappers' story was the tragedy, and not the story of the princess's love? After all, the article might not have conveyed the spouse-swappers' true feelings, or maybe they hadn't told the reporters what they really felt.

Was it really possible that the children—living right next door to their former apartments, across the street from their old houses—would find the father-swap "a pleasure?" Was it really "better for the children?"

Takako couldn't believe it.

Spouse-swapping wasn't the kind of thing a person could do ordinarily, of course—certainly anyone predisposed to tragedy would have a hard time going through with it. To think that all four people, two married couples, had felt the same way—it must be incredibly rare. Indeed, it was precisely because it was so rare that the topic had been taken up as international news in the first place, and why the Japanese newspaper had chosen the story for "This Country, That Country."

The eight Swedes and Americans must have been either frivolous pleasure-seekers or lawless rebels—it was inconceivable that they had thought very deeply about what they were doing—and they must have exchanged spouses in the spirit of comedy. Takako was sure this was the case.

Even so, there was no denying that something utterly improbable had happened, really happened, twice within the past three or four days—even if it had happened in distant countries.

Takako, twenty-nine, was certainly well aware that things one might assume to be impossible sometimes do happen in the world. People can bring themselves to do anything at all, for any reason whatsoever. You can never tell what a person might do.

But she could never do it herself. She never would.

"I suppose all you really need are four people who feel the same way about it—four people out of an uncountable number—and then it's possible, isn't it?" she muttered, forcing herself to laugh.

No, it wasn't amazing. It would cause no very terrible inconvenience, it wasn't a crime. And yet "all you really need" was not something one could honestly say.

Takako decided that it would be best not to show the article to her husband, Hirata.

Hirata had skimmed the newspaper before he left that morning—it was unlikely that he would look through it again, or that if he did he would happen across "This Country, That Country." Even supposing that he did read the column, he'd probably think it was just an amusing topic, a nice little snippet of a story. He'd probably be more interested in the story about Princess Margaret and Group Captain Townshend, the story of the mound of stones.

Still, Takako decided to put the newspaper away somewhere where he wouldn't see it.

2.

Impatiently, Takako tried to jam the newspaper that contained the story in near the bottom of the stack she was making in the corner of the closet, but she couldn't make it go in.

She pictured herself squatting there, disgracefully posed, and suddenly found herself besieged by sinful thoughts.

Hiding the newspaper was not her only sin.

Sliding the closet door shut and turning toward the room, Takako was startled by the vivid shadow of a tree on the paper-paneled door. A bright autumn sun was shining outside.

She went out into the garden.

The shadow she had seen was being cast by a holly. The tree was larger than most hollies and was the only one in their garden that really looked like a tree.

The holly was speckled with tiny white flowers. Though it was plainly visible from the sitting room,

Takako couldn't remember when the flowers had started to bloom. It was strange that she couldn't remember.

And now the flowers were falling—the black earth in the tree's shadow looked white.

Takako picked up three or four of the small flowers and held them in the palm of her hand. Each blossom had four round, softly curved petals. The stamens were long.

Hirata might have noticed the flowers on the holly, but of course he would know nothing at all of the delicate form of the individual blossoms. So far neither Takako nor Hirata had mentioned the flowers this autumn.

Hearing the swish of *kumazasa* bamboo in the garden of the house next door, Takako called out, "Ricky, Ricky!"

A brown mutt put its head through a hole in the bottom of the bamboo fence. Takako could see from the movements of his head that the dog was vigorously wagging his tail, but he stayed right where he was and didn't come through to their garden.

"Ricky, has Mr. Chiba gone out?"

Takako spoke so that Chiba would hear if he were home.

Chiba had named Ricky after the pro-wrestler Rikidōsan, of course—his nickname was "Ricky."

"I bet Rikidōsan would be angry if he found out," Takako had once said to Chiba.

"I doubt if he ever will. And even if he did, he'd

probably just take it as a sign of his popularity and laugh it off. There really aren't many good names for dogs, plus he's a male, and he's a guard dog, so don't you think it's right for him to be named after someone strong? Though once when I was walking in town I heard someone holler out 'Ricky, Ricky!' and when I looked it was this little terrier. Well, I thought, so other people are using it, too—trendy people."

"It's a nice name. It has a nice sound."

"You must have had the same experience yourself. You hear someone call out 'Takako' somewhere and you spin around. . . ."

"Yes. There aren't too many names for women, either."

"Ricky's neck, the way it's so long . . . it kind of looks like yours. At least I think so."

"Are you saying that this dog looks like me?"

Takako thought she might laugh, but she didn't. She wasn't angry—it wasn't that, exactly. But to think that Chiba had looked at his pet dog's neck, thinking of hers! Her cheeks reddened.

Takako knew that her neck was slimmer and more shapely than those of most Japanese women— her friends had pointed this out to her since the time she was in grade school. Her neck had remained beautiful even after her marriage, never growing fleshy or unpleasant to look at. Hirata was also aware of the beauty of her neck. He sometimes pushed her jaw upward with his forehead and kissed it. Takako was so used to this that it no longer even tickled her.

Yet whenever young Fujiki brushed her neck with his lips, Takako felt so ticklish that she leapt up and leaned away, shrieking.

This difference surprised Takako, even frightened her. It was hard for her to believe that she responded to Fujiki the way she did solely as a result of his shyness, the softness with which he touched his lips to her neck.

"Ricky, come here."

Takako called the dog again. But he kept standing there, his head and neck poking through the bamboo fence.

The fence was very old—it had been built before Takako's arrival at the house. Thick pieces of bamboo had been split in half and lined up with their insides facing the Hiratas' side, a fact that suggested it was someone from the Chibas' house who had built it. The akebi vine that had climbed the bamboo on the Chibas' side sometimes dropped its shriveled berries into the Hiratas' garden.

Chiba had still been single when Takako arrived as Hirata's bride. He was living with his mother and his younger sister, and there had been a pretty maid. The sister married and moved out soon after Takako arrived. Two years later Chiba's mother died.

The Hiratas were invited to the sister's wedding reception, and they had also attended the mother's funeral. And Chiba had been at the Hiratas' wedding reception, although Takako, the bride, had not seen him.

Two or three days after the Hiratas returned from their honeymoon—it was a Sunday—Chiba had called down from the second floor of his house,

"Hirata, Hirata."

"Yes?"

Takako walked out into the hall. It was the first time she had heard the name "Hirata" called like that, from a distance, and she'd taken it to refer to herself. It was also the first time she saw her neighbor's face—Chiba's face.

Chiba seemed somewhat taken aback to see her come out.

"Oh—I'm sorry to be shouting down at you like that. . . . I've had a pheasant sent up from the country, and I thought I'd send it over as a wedding present. I mean, if you'd like it. . . ."

"Well—yes."

Then Takako's face became red and she went back inside, making a gesture as if to say, wait a moment please! There was probably no need for her to ask her husband, since Chiba had said he would give it to them, yet . . .

Takako's chest was pounding. Chiba's voice was ringing inside her.

Hirata was delighted with the message she brought.

"Do you know how to prepare a pheasant?"

"No! Oh, no—I'd be afraid to. I couldn't. And—what does he do?"

"Mr. Chiba? He's an architect, works for a construction company. He graduated from a private university, but I hear he's got talent."

Hirata stepped outside, and a moment later he and Chiba were tossing words back and forth between the second floor of Chiba's house and the first floor of the Hiratas', saying—My wife doesn't know how to cook pheasant—Shall we see if they'll fix it up for us at the poultry store?—And of course you'll have to join us, Mr. Chiba—Shall we have it in broth, family style?

The Chibas' house had two stories, the Hiratas' house only one. As the second son, Hirata had been obliged to leave his parents' house and move into a house of his own when he married. For this reason, the Chiba house was generally more impressive.

"Will Mr. Chiba be joining us?"

Takako had been listening inside and knew the answer to this question, but she asked it anyway.

"Yeah, I thought I'd ask him over. His mother and sister helped out a lot when my mother was here. Before you came."

All this had been before Ricky's arrival at the Chiba house.

Ricky had come to the Chibas' as a fully grown adult. Surely—surely Chiba couldn't have seen him somewhere and asked to have him because his long neck reminded him of her? No—that would be going too far.

Ricky kept standing with his head and neck—that neck, so slim for a mutt's—poking through the hole in the bottom of the bamboo fence. He still made no attempt to step through it.

"Has Ichiko gone out, too? Are you watching the house, Ricky?"

Ricky was staring at her.

Takako didn't have to ask Ricky to know that Ichiko, Chiba's wife, was out. The ceaseless stream of voluble chatter that Ichiko addressed to her three-year-old daughter, Yōko, and to their maid, had stopped, and the radio, which Ichiko left playing from morning to night, was silent.

The Chibas had a television, too. Chiba had designed a summer house in Atami for the owner of a bicycle factory, and when he finished he was given a woman's bicycle and a television as a gift.

Ichiko had once seen Takahashi Keizō, a popular announcer on NHK, in the hallway of a kabuki theater. She had greeted him by mistake, bowing. "Why, if it isn't Mr. Takahashi!" she had said. And Takahashi had returned her greeting very smartly, very professionally. "Well, you always see him on television, right?—so I figured he must be someone I knew. Afterward it all seemed so funny. . . ."

For a time, Ichiko had told this story to everyone she met. Takako had heard her telling it, and heard her laughing, next door.

She often saw Ichiko riding by on her bicycle, wearing a pair of ski pants that her husband had worn in his student days, with the three-year-old Yōko strapped in behind her. Ichiko had chosen the name Yōko because it fit her own simple requirements for a name. The *ba* in Chiba and the *yō* in Yōko were both written with the same character, so the four characters in the name could be divided into two more or less similar halves. Ichiko said this was good; she was proud that this also was true of her own name.

Ichiko had once said to Takako, "I've married into a family with lucky names."

The Chibas' maid was a woman in her forties. She was a war-widow, a quiet woman and good worker. She walked about sunk in silence, picking up the things that Ichiko left scattered throughout the house. With a husband named Chiba and a maid named Taneko, Ichiko felt sure she must be blessed with good luck.★

Even though Ichiko was generally unconcerned with subtleties and nuances, Takako thought she must have sensed her affection for Chiba, that Ichiko must know she had been attracted to him

---

★"Chiba" contains a character meaning "leaf," and "Taneko" contains one meaning "seed." Tr.

since soon after she arrived at Hirata's house. Yet Ichiko put no distance between them, and never seemed embarrassed.

For example, Ichiko thought nothing of it if Takako saw her when she was changing clothes upstairs. Sometimes she would walk to the window, still naked, and start talking to Takako.

The living room downstairs and the bedroom on the second floor of the Chibas' house were furnished western-style. The bedroom seemed originally to have been a Japanese-style room with a tatami floor, which they had redecorated as a western room. The chest of drawers where Chiba and Ichiko kept their clothes was built into the wall of their bedroom. If Takako was careless enough to look, she could see both the Chibas changing clothes from inside her own house. The Chibas' bedroom was at the back of the house, nearly hidden, but occasionally in summer Takako could see through the window that Chiba had put a large board on top of the double bed and spread blueprints out across it.

Takako sometimes wondered, somewhat strangely, if the fact that Chiba used the double bed as a desk meant that he was satisfied with Ichiko's body. She herself had never been in a double bed. What was it like in a double bed, she wondered, in that other country.

Hirata and Fujiki were also as different as this country and that country. Hirata disliked it when Takako moved—with Fujiki, though, she had been the one to move from the beginning. Of course this might have been due to the fact that Fujiki was still a student, so much younger than she.

"I feel so humiliated. I'm such an animal." Takako often said such things when they were finished, covering her face, her wide-open eyes. And Fujiki would rejoice in the knowledge that Takako felt so passionately toward him.

Takako thrashed madly in bed to paralyze her sinful thoughts.

She felt that if Chiba were her partner she could be docile, that her eyes would brim with tears of joy, though the sin would be the same.

Whenever it seemed that she was about to cry out, Hirata put a finger into Takako's mouth, making her bite it.

It wasn't exactly true that in forcing Takako to restrain herself Hirata satisfied her less than Fujiki. Indeed, when she was with Hirata, Takako sank into a sweet and natural sleep—a sleep that never came with Fujiki. Neither was it quite true that Takako desired in Fujiki a spontaneity that Hirata lacked. She always felt weary when she and Fujiki were through, and mixed in with her weariness were loneliness and regret. It wasn't simply that Fujiki was a person whose existence she had hidden from her husband; perhaps her body had been trained somehow, somehow made accustomed to Hirata's.

Ultimately, the difference between the person Takako was with Hirata and the person she was with Fujiki was so great that she found herself unable to believe that the two women were the same. She was astonished to discover those two different women existing within herself—their reality must mean that she had become immoral, or that she had passed beyond morality.

The discovery that two women existed within her—occurring as it did after she had committed an immoral act—was strange, and of course it caused Takako pain. Yet at the same time, she found it giving life to another secret desire, the desire for a third man.

Takako had made up her mind not to pursue Chiba—she had decided that this was one thing she simply could not do. But since she had become involved with Fujiki, she had come to think that it was not so impossible to get involved with him after all.

Actually, Takako did not consider Chiba "a third man." He was "the first man."

Her involvement with Fujiki had been a "mistake," she thought, as if she had been caught up suddenly in a whirlwind, lost control, and Takako regretted that it had happened. At the time she had felt that she was betraying Hirata, yet even then, in the very depths of that feeling, lurked a sense that she had betrayed Chiba. Takako found this strange, but somehow the very strangeness of it set fires burning within her.

Takako looked away from Ricky, and he pulled his head back through the hole in the fence.

The curtains of the Chibas' second-floor bed-room were closed.

Takako walked in the direction the dog had gone. Peering through a gap in the fence, she could see that the well-trimmed hedge of sasanqua that separated the front door from the side door of the Chibas' house was covered with blossoms. The blooms were past their prime, and some had dropped—they formed a layer on the ground.

In one part of the garden, over on the other side of the hedge, cosmos and yellow and white chrysanthemums and some variety of purple flower were blooming riotously, like a clump of weeds. The roots and seeds from last year or the year before must have sent out shoots, shoots which now blossomed at random, uncontrolled. It looked as if the soil had originally been built up into a neat circular mound, but now the various flowers had grown randomly in all directions, drooping over, so that the outline of the flower bed had disappeared. The leaves of several of the flowers were starting to wither.

Still, Takako had never seen such a fantastic variety of different flowers in the Chibas' garden, blooming the way they were now. Her heart fluttered.

With Chiba and Ichiko both away from home, it seemed impossible, somehow, that the flowers should be standing so still in the autumn sun.

"I shouldn't be here."

Takako muttered this, and stepped away from the fence.

The holly flowers she had picked up earlier still lay in the palm of her hand. She was about to toss them away, but reconsidered and took them with her into the sitting room. She lay the tiny white flowers on the *kotatsu*, a *hori-gotatsu* into which they had inserted an electric coil.

She heard a sound, as if the front door were being eased open.

"Hello? Takako?"

It was Fujiki, whispering. Takako leapt up in confusion and went out to meet him.

"What are you doing here?—You can't come here! You must be crazy! What on earth are you thinking of?"

"I know, I understand—I didn't want to come, but . . ."

"Then why did you come?"

"I had to. You haven't come to see me for over a month."

"But—look, it's bad for me if you come here. We'll meet tomorrow."

"There's no one else here, is there? I made sure before I came in."

"What?"

"I was watching for a while outside the gate."

"You were—I can't believe it. What a hideous thing to do."

Takako's face was red, and she was nearly shaking.

"Even if there's no one here, there are people next door. Didn't it occur to you that the people next door will see us?"

"Are the people next door likely to be watching, checking to see who comes here? Besides, even if they did see us, they wouldn't have the foggiest idea who I am or where I come from."

"No—you're wrong."

"Have you told someone next door about me?"

"No, it's not that! It's just that the people next door—I've . . ." Takako couldn't go on. "You scare me. If you're going to start coming here like this, it's over between us. The end! I've been thinking I made a mistake anyway."

"Okay, don't shout—calm down. . . ."

"No—I'll shout if I want to! How dare you tell me to calm down!"

"But what's happened?" Fujiki asked, starting to take off his shoes.

"Stop it, stop it!" Takako said, backing away. "It doesn't have to be tomorrow—we can meet today. I'll come see you, I promise, just as long as it's somewhere else—please, will you leave? All right? Please, just get out."

The terror she felt at Fujiki's arrival had to do with her husband, of course, but a sudden spark of feeling told her that it had to do with Chiba, too. Her loss of control was his fault as well as Hirata's. The Hiratas' plank fence was full of holes, and one could peep through it anywhere if one wanted to.

Even though the Chibas were both out, and their maid, Taneko, was the only one home, it seemed as though Chiba were home, as though his eyes were fixed on her and Fujiki.

The thought that Fujiki had been peeping in at her as she peeped in at the flowers in the Chibas' garden filled her with shame—shame that was at the same time hatred for Fujiki.

Takako still hadn't calmed down. Fujiki stepped hurriedly up into the house, put his arms low around her back, and hugged her to him.

"Let me go! Let me go! What have I ever done to you? Forgive me. . . . Oh—I hate you!" Takako's words came in bursts. Her face lost its color.

"Hey—hey."

Fujiki shook Takako as if to call her back.

3.

Hirata stood on a street in Yotsuya Shio-chō, waiting for an empty taxi.

He was neither particularly cautious nor particularly nervous, but it was his custom when hailing taxis to get in only after he had inspected both the vehicle and its driver.

On this occasion, as on others, he hailed a taxi with a pleasant-looking old man in the driver's seat.

"Namiki Street in Ginza."

When they arrived at Yotsuya, the driver asked vacantly, "Which way do you want to go?"

"What? Either way is fine."

"Should I go right?"

Hanzōmon was straight ahead, Akasaka was off to the right. Coming as they had from Shio-chō, it made the most sense to continue straight through the intersection, but there was hardly any difference between that and going right, heading into Ginza from Akasaka.

The taxi turned right.

Like most old men, the driver was quiet. Hirata relaxed and sat gazing out at four or five swans floating in a corner of the moat, close to Benkei Bridge.

"They seem to have released swans here, too—do you know how long they've been here? This is the first time I've seen them," he said to the driver, but there was no response.

Ten or so rather large waterfowl came flapping down. They were wild birds, their wings not clipped like the swans'. Hirata felt how close winter was, when one often saw flocks of waterfowl in the moat.

At the same time, he basked in the warmth of the pleasant autumn sun, some rays of which found their way even into the peaceful interior of the car. The taxi climbed the slope toward the National Diet.

At the fork in the road that led toward the Diet in one direction and toward Mittakuzaka in the other, the driver asked another stupid question.

"Which way do you want to go?"

"Right," Hirata said, noticing now that there was something a little odd about him.

Looking again at the driver's face from his seat in the back, Hirata saw that the man he had taken for a rather pleasant fellow when he entered the car was really a senile old fool.

Hirata spat out a question as they passed the Diet.

"Are you telling me you don't know the streets in the middle of Tokyo? Where are you from?"

"Well, I've been out in the country for fifteen years. . . ."

"What?"

Hirata felt himself in danger and slid over, positioning himself directly behind the driver.

"Tokyo's streets have changed a lot these fifteen years. . . . I drove for a long time in Tokyo before. But roads in the country are hard too, you know. Those roads in the mountains. . . ." the man said, lifting his left arm from the steering wheel and using it to describe what must have been the slope of a mountain road. He turned his hand toward the front of the car and slowly lifted it up.

"Hey, watch it," Hirata shouted.

"I just got back to Tokyo a month ago. I never know where I am."

"Well, drive slowly."

A large billboard on the street in front of the Metropolitan Police Board caught Hirata's eye: "Yesterday's traffic accidents: 3 dead, 25 injured."

"I go right?"

"Yeah."

Hirata had spoken without thinking. If they continued straight on, just before them—wasn't that the Hibiya intersection, Ginza yon-chōme? To ask the way at this point was really too much. Was the man a fool?

The taxi seemed to stagger as it made the turn. It occurred to Hirata that it might be safest to get out of Kasumigaseki anyway, to go on into Hibiya Park.

It was impossible to say how many cars had passed them.

Hirata began to finger the large mole on the left side of his jaw. He did this whenever he was irritated.

He knew that taxi drivers were required to earn a certain amount of money every day—this was why they took out-of-the-way routes. If a driver didn't earn what he was supposed to he would be assigned a bad car. Or worse yet, he might lose his car altogether—he could end up working as a "sub," a driver who fills in when others are on vacation.

Hirata had also heard that Tokyo's taxi drivers overworked the nerves in their eyes, so that by the time they reached middle age their eyes no longer functioned properly. They generally had to retire pretty early.

A driver as old as this one wouldn't make it in Tokyo.

"Wasn't it easier for you driving in the country, crossing mountains?"

"No, not really."

Somehow they got lost in Hibiya Park and the car ended up in front of the public hall. The driver looked frightened and turned away from the hall into the shadow of some trees, a place that was not a road.

At last they made it out of the park.

"Keep going straight. . . Just keep going straight. . . ." Hirata said each time they approached an intersection, glancing nervously to the left and right.

To think that he had chosen this taxi himself, this senile driver—Hirata started thinking that his luck must have run out. Mirroring his unease, his broad round face grew distorted.

Hirata, a fundamentally light-hearted and selfish man, was sometimes overcome by this feeling of unease. It had been happening ever since he had grown suspicious of Takako.

From the very first—soon after he had moved into his present house and married her—Hirata had been fond of Chiba, had been attracted by his personality, and though Chiba was five years his junior he had looked up to him. He had been pleased to see that Takako liked Chiba, too—at least until recently. Now he found himself suspicious of them.

No—it wasn't *them* he was uneasy about. Chiba was entirely without guilt. It was Takako he suspected.

At times Takako's eyes would refuse to settle on anything, and she would seem to be gazing off into the distance. The glint in her eyes had changed.

After the pheasant dinner, over which he'd introduced Takako to Chiba, Hirata had said brightly,

"Well, what do you think—he's a nice guy, isn't he? Maybe you were really meant to go next door but you got the houses mixed up and married me— what do you think?" The question had been an expression of his pride and pleasure at having made Takako his, but in retrospect it seemed to have been an unlucky thing to say.

At the time, Takako had only asked how old Chiba was.

"I think maybe five or six years younger than me. But he's smart."

"He's young, isn't he?"

Then Ichiko married Chiba and moved in. Not long after the Hiratas had become acquainted with her, Takako said to Hirata, "She's not at all like Mr. Chiba, is she? I'm a little disappointed."

"No—but they're the kind of couple where everything that doesn't seem to match in them ends up matching. Call it heaven's dispensation, or something like that—after all, that's what marriage is. You might think it'd be better if a woman more like you had married him, but actually that wouldn't work out at all. Yes, I think he's done all right," Hirata said, nodding in agreement with himself.

Hirata had underestimated Takako, assuming that as a woman she was unable to understand the allure of Ichiko's physique. He would say nothing of this to her, for while he kept tight control over her, he also recognized that she was superior to him in certain ways. And besides, he never discussed sexual matters with her. He thought he saw pleasure in the mix of things that filled Takako and flowed from her as he drew her gradually on, a little and then a little more, in contented silence. He assumed that she herself found satisfaction in polite self-control, that the key to a long future as husband and wife lay in her self-control.

But now, suddenly, Takako seemed to have changed. Sometimes she would reveal an aspect of this change in some careless act, and Hirata would feel his body tightening.

Another man had laid his hands on Takako—he could conceive of no other explanation. And he could imagine only that this man was Chiba. He knew nothing of Fujiki.

Suddenly it occurred to him that he might have ended up in this taxi because Takako was secretly wishing that he would die in an accident—it was a morbid and unjust suspicion, lacking in dignity— and it seemed as though his body were going numb from the feet up.

"This is fine—let me out here."

The taxi had arrived in Ginza.

"You can just stop here. The roads in Ginza are all one way, it'll be too much of a hassle. I'm sure you don't know the streets."

The elderly driver attempted to pull over, but in the swarm of cars he was unable to find a stopping place. He continued unsteadily out into the intersection. He attempted to turn right and slipped along surrounded by the noise of the horns of the cars behind them for nearly sixty feet.

Looking with relief at the street outside, Hirata was amazed to see that it was Namiki Street.

"How much?"

"One hundred fifty yen."

He was surprised at the cheap fare. Was the meter broken? Or had the driver misread it?

He groped about in his pants pocket for change, but perhaps because his body was still numb from his sense of the danger he had been in, he couldn't find any. He was still feeling unsettled when a young woman wearing pants approached the car window.

"Hey, it's illegal to stop here. You'll get in trouble."

The young woman spoke with a languid gentleness. Somehow Hirata had the impression that she was smiling, but neither her eyes nor any other part of her face actually smiled.

Still, the young woman's smile helped him. His encounter there with the particular sense of harmony and good will that women emit caused his suspicions of Takako to vanish.

"Yes, yes—of course."

Hirata answered on behalf of the driver and hurriedly climbed out of the taxi.

Having walked on a little way, he turned back, and then turned back again. The young woman was watching him walk off, looking as though she had something to say.

She was wearing a tight coal-black sweater over deep brown pants, and her tanned face was not made-up. Her eyes were narrow, and her eyelids were entirely unwrinkled. Her slender, rather masculine figure looked as though it was disintegrating from negligence, and he found this peculiarly charming.

Hirata guessed that she was a street girl. This too was new and fresh.

Dictatorial in his dealings with Takako but at the same time powerfully restrained himself, Hirata felt for the first time an urge to rebel. He walked to the corner of the street and then walked back.

The young woman was standing in the same place.

"You seem like a nice girl. What are you doing hanging around here?"

"I do the cigarettes at the pachinko parlor."

"Ah."

The woman's answer was extremely concise. Hirata nodded, as though he were responding to its echo, then smiled as if to say, "I see."

They were standing before a large pachinko parlor.

Hirata had never been inside a pachinko parlor, so he wasn't really sure what to make of the woman's reply. He wondered vaguely whether she might exchange cigarettes that customers won for money.

"Pleased to meet you."

The woman bowed her head lightly, perhaps because Hirata would not leave.

"I work in the advertising section of a pharmaceutical company in Nihon-bashi."

He had said too much.

"Pharmaceuticals?"

"Yes."

Though technically he was in the advertising section, Hirata actually worked as a salesman—he visited stores. Different companies sold more or less the same faddish medicines, and most consumers knew almost nothing about even common things like cold medicines and multi-vitamins, so most people made their purchases according to clerks' recommendations. Hirata's job was to visit stores and make friends with the clerks.

Just as he was about to say something else to the young woman, he noticed that his neighbor, Ichiko, was approaching.

Ichiko came within inches of Hirata, who had been waiting to greet her, but kept going—she didn't even notice him. Her face as she passed was set in the ill-tempered expression of a woman walking alone.

"People really are bizarre, aren't they? You never know when someone might be watching you without you noticing—it could happen anywhere, no matter what you're doing," Hirata said.

Ichiko was wearing a winter coat with low shoulders and a high hemline—the kind that was fashionable that year. She was short, and her figure as he saw it now from behind resembled a bag of rice.

In fact, winter coats like bags of rice were walking around all over.

Hirata felt pleased and ran after Ichiko. He tapped her suddenly on the shoulder.

"Well—if it isn't Mr. Hirata! You must be trying to startle me."

"You walked right by me."

"Did I? How awful of me."

"Are you out shopping alone?"

"No. I just came to pick up this coat."

"Well." Hirata leaned back and looked at Ichiko's coat again. "It's nice, isn't it? Shall we go sit down somewhere, maybe have some tea?"

"To tell the truth, rather than tea, I'm dreadfully hungry. A woman can't go into a restaurant alone, you know."

"You ought to have asked Takako to come along."

"Yes, I should have. But I left this morning, I had to leave Yōko with my mother. If you don't mind perhaps I'll ask her along next time."

"I hope you will."

Ichiko seemed completely unsuspicious of Chiba and Takako.

"I hope you'll let me pay," said Ichiko brightly. "There's a marvelous French restaurant that Chiba goes to all the time—shall we go there?"

"I'd love to—though the timing seems a bit odd, doesn't it?" Hirata's round face filled with a smile. He hardly knew how to account for the pleasure that came bubbling up inside of him. "Certainly a lucky meeting, don't you think?"

4.

"Mrs. Hirata! Takako!" Ichiko shouted down in a hearty, pleasant voice from the window of the Chibas' second-floor bedroom. She had slid the glass window open with a fair amount of force before calling out, and she assumed that Takako would have heard the noise it had made next door.

Ichiko very rarely called Takako from upstairs. On most occasions she walked to the old bamboo fence and called her from there—though this, too, was something only Ichiko did. Takako herself never shouted at Ichiko like this. If she had something to discuss, she would go around to the gate of Ichiko's house and then walk to the front door.

"Takako! Takako!"

Ichiko called again, brushing her hair.

Takako opened the paper-paneled door, darkened by the shadow of the holly, and looked up at Ichiko.

"What are you doing?" Ichiko asked.

"Nothing much."

"Were you reading the newspaper?"

"Yes," said Takako, squinting.

"Good guess, don't you think?"

Ichiko shrugged and laughed. Her laugh echoed brightly.

Takako's pale face was bathed in sunlight—it looked soft to Ichiko, who was nearsighted. She felt as though Takako were gazing up at her, here on the second floor, in pure admiration. What a lovely, innocent person she was.

"Takako, would you like to come and talk a bit? If you're free?"

"Yes, I'd like that."

"Let's have a nice talk, okay?"

Ichiko was staring directly at her, so Takako laid a hand on the collar of her cardigan and turned away. Ichiko was really only looking at Takako's face, but it must have seemed that her gaze penetrated into the farthest reaches of the sitting room. Ichiko's near-sightedness wasn't really very bad—Takako had never seen her wearing glasses and had no idea she wore them.

"This weather is so nice I can hardly sit still. I was thinking I might go somewhere, so I came up here to get myself ready, and then all of a sudden I found myself wanting to talk to you. You'll come over?" said Ichiko.

"All right."

"You know, I ran into your husband a few days ago—in Ginza. Did he tell you we'd met?"

"No."

"He didn't?"

Ichiko had assumed that Hirata would mention their meeting to Takako when he got back, almost as a matter of course.

"We went to a French restaurant that my husband likes. I'd been trying to decide what to do, since I couldn't eat alone, and the timing was perfect."

"You must have treated him, then—I hope he didn't cost you too much. I'd have thanked you sooner, only he didn't say anything."

"Oh, you don't need to thank me," Ichiko said, laughing brightly.

Takako went around to the gate of the Chibas' house and stood waiting at the front door, but Ichiko called down from upstairs, "Takako, you can just come in." Takako waited a little, but Ichiko didn't come down.

"Taneko!—Taneko!" Ichiko called the maid. "Show Takako in, please."

Takako heard Ichiko shouting for Taneko all the time. Because Ichiko's voice was so loud, Takako often picked up bits and pieces of information about goings-on in the Chiba household. Some days she even knew what they were having for dinner.

The story of Ichiko's fight with the owner of the tofu store was famous in the neighborhood. Ichiko had insisted that some tofu her maid had brought back was spoiled, and flew off on her bicycle to complain. The argument grew so violent that a crowd formed outside the store.

The tofu store's owner got excited and shook a knife at Ichiko. She filed a complaint at a police box on her way home. A policeman led the owner of the store to her house and made him apologize.

"My, you've got a nice house," the store owner said.

Judging from Ichiko's indifference to the crowd, and from the rudeness of her shouts, he must have expected quite a different kind of house.

"When he brought out that knife!—To tell the truth, I was so terrified I thought I'd start trembling. But I stood my ground!" Ichiko said afterward, as if it were all a joke. "But you know, thinking back on it now, it must have been a tofu knife, right?"

Her audience always laughed, so Ichiko would start laughing first.

Takako had even heard Ichiko telling the story to the delivery boy from the fish store, and to the man who collected money for the newspaper. She would be in her sitting room and would hear Ichiko telling the story in the kitchen of her own house.

It wasn't at all uncommon for Ichiko to answer salesmen who came to the kitchen door by shouting out to them from inside the house, and then to keep them waiting for a long time. She left them alone for so long that Takako worried something might be stolen.

Now, even though Ichiko had asked her over, there was nothing to suggest that she was on her way down to meet Takako.

The maid, Taneko, led Takako to the living room. Takako stopped at the door, surprised.

"Isn't this Mr. Chiba's office?"

"Well, it's where he works."

"Is it all right for me to be here?"

"Yes, of course—go right in," said Taneko.

Both the table for receiving guests in the middle of the room and the larger table by the window were covered with western books and architecture magazines. Blueprints lay spread across the table by the window, and on top of these lay various drawing tools. Architectural photographs were stuck up on all the walls, so many that they overlapped in places.

Though she lived next door, Takako had only rarely entered the Chibas' house before. She and Ichiko would stand talking over the bamboo fence, or else say what they needed to say at a front door or verandah. And on the rare occasions when she had come inside, she'd usually been shown into the sitting room. This was the first time she had been brought into the living room.

Chiba went in to the construction company's office to work, and he met with customers there, so it was unlikely that he used this living room very often. He used it mostly as a study.

Takako remained standing even after the maid had left, still unable to relax. She walked around looking at the photographs on the wall. Most of them showed buildings by western architects, but a few seemed to be buildings Chiba had designed. There was a wine bottle on the round table, and a few glasses.

Ichiko was still at the vanity in the bedroom, arranging her hair. She didn't usually bother with it very much, so it took time to get it into the shape she wanted. She would have to change clothes as well. Most mornings Ichiko would simply wash her face and put a bathrobe on over her pajamas—she seldom put on makeup—and then see Chiba off to work. She would stay dressed like this until the afternoon. Now that autumn was under way and it was growing chilly, she'd change into a pair of Chiba's old pants and an old sweater.

But earlier, gazing down at Takako from the bedroom window, Ichiko had found her so extremely beautiful that she had seemed almost sacred. It was for this reason that she planned to change her sweater for a nicer one, tossing all her sweaters out onto the double bed.

Ichiko touched up her eyebrows with an eyebrow pencil. Her eyebrows were so full naturally that she could hardly add to them.

Then she went down to the living room.

"Oh, I'm sorry—I didn't mean the maid to show you into this mess. Shall we go into the sitting room?" Ichiko asked, but then she sat down in front of Takako, crossed her legs, and lifted the lid of the cigarette box on the table. The box commemorated the completion of a construction project.

"Doesn't Mr. Chiba work in this room?"

"That's right. He was up working in here until about three-thirty last night. Things seem to come together better for him at night."

"It must be rough for you, too."

"Oh no—I go to bed. And once I'm asleep I don't wake up again until the next morning."

Ichiko noticed that Takako was uncomfortable—no doubt she sensed something of Chiba in the room. The thought filled Ichiko with a prankish awareness of her own superiority to Takako. She was extremely sensitive in matters relating to men.

Though Ichiko was often seen riding about on her bicycle in a pair of her husband's old pants, in certain ways she was not what she appeared to be: after several years of marriage, she still grew weak when her husband kissed her, and had trouble standing. Chiba would reach hurriedly to support her—it seemed that he was happy with her.

"You and I are neighbors, but somehow we almost never have a chance to sit and talk," said Ichiko.

"It's true, isn't it. . . ." said Takako, lowering her eyes. Then, "Does Mr. Chiba drink?"

"You mean this?" Ichiko placed her hand on the wine bottle. "It's pretty, isn't it?"

"I've been looking at it, thinking how beautiful it is. The color is like an emerald's."

"It's Czechoslovakian cut glass. I've been told that it's absolutely impossible to produce this color in Japan. My husband keeps it as a memento of his father, who died when he was eleven. Apparently he loved this bottle—it was very important to him. Whenever he was in a bad mood he would take it out and look at it, and this would make him feel better."

"Yes, it really does make you feel good, looking at it."

"Apparently if the kids even went near their father when he had this out he would yell at them—it was so bad Chiba used to think his father liked the cut glass better than he liked his own kids. His father didn't even want to use it—he said he was going to put good saké in it and drink from it when he got old. He was looking forward to that, but in the end he died without ever having used it."

"Really?"

"Chiba's mother sold all sorts of things to put him through college, but this she wouldn't sell, it's the only thing she kept as a memento. After his mother died, Chiba started using it. I'm sure it reminds him of his parents. I handle it as gently as I can myself, and I try to touch it as little as possible. But since it's so fragile it'll probably break eventually."

Takako had been staring at the bottle all the time Ichiko spoke. The clearness and the luster of the bright emerald green had the beauty of a jewel. Something clean filled her chest, and she felt relaxed. The bottle was rectangular, shaped like a *hentsubo*, and the neck was a rectangle, too, stopped with a large cork. The glass was dotted with round depressions, like a pattern of water drops. In each of these round depressions the objects on the far side of the bottle could be seen—tiny, tiny reflections. The same objects were visible in the depressions on the opposite face of the bottle, so it seemed that infinite numbers of tiny objects were standing there, all lined up in rows. The deep brown chinaware cigarette box, the white cigarettes within the box, the color pictures in the architecture magazines—all this hovered miniaturized and multiplied within the emerald light. If she lowered her head slightly and looked at the bottle, she could see Ichiko's lilac sweater, her breasts, miniaturized and multiplied.

"Chiba only puts good wines in it. It looks like there's a little left at the bottom, doesn't it?" said Ichiko, moving her face closer to the bottle. Then she noticed that Takako's eyes were filled with tears. She sat back up in surprise and looked down at her face.

"Are you crying, Takako?"

She had startled Takako.

"It's so beautiful I cry," Takako said, blushing faintly. When she lifted her face Ichiko was struck by the beauty of her long, thin neck.

"You're so sensitive, Takako," said Ichiko with genuine goodwill, feeling that she would like her husband to see Takako as she was right now, with just this expression on her face. This expression, from looking at cut glass! There was no jealousy in her.

"Once when your husband was here he and Chiba drank together from this bottle. He didn't mention it?"

"No. But Hirata has no taste for this sort of thing. . . . I doubt he'd even notice," said Takako. She changed the subject. "You said you had been to eat in Ginza, didn't you?"

"You know, my taste isn't very refined either. I find Mr. Hirata interesting. I like him."

Takako was looking around the room. Every time her eyes fell on one of the old architectural photographs on the wall she felt ashamed of her own husband's job, a medicine salesman who ran the company's advertising department.

"We had western food, it was a lot of fun. 'Takako doesn't seem to be satisfied with me,' he was saying. Well, Chiba's the same way. A woman like me—I can't really be intimate with him. I cuddle up to him and so on, but to tell the truth I sometimes think that you ought to have come here, and I ought to have gone to Mr. Hirata's—the two of us should have been reversed. There's nothing we can do about it now, of course, but. . . . And if that's the case with us, even with us—the Hiratas and the Chibas, two couples living right next door—I really think most married couples must be mismatched. Yet there's no going back."

Ichiko knew how deeply Takako herself believed what she was saying, and felt a certain measure of security in knowing that everything was already settled. It was this that made it possible for her to speak as she did. Her wheat-colored cheeks seemed full of life.

"That can't be true. The two of you always get along so well together. I envy you," said Takako.

"Chiba makes sure we get along. There are all sorts of things about me that he'd like to complain about, but—I suppose in certain ways that makes life easier."

A dog whined at the gate.

"He's home!" Takako was the first to speak.

"Indeed he is. How clever of you to know."

When Chiba's footsteps drew near the gate, Ricky would become excited and start whining on the other side. This whine was different from his bark. Takako always listened for it from next door, and each time she heard it she thought: Chiba's home!

Ricky leapt at Chiba's pants and followed him into the entryway.

Takako left the living room with Ichiko and stood behind her, her shoulders stiff.

"What's wrong?" Ichiko asked him.

"Oh, I forgot something."

Something unpleasant seemed to have happened to Chiba at work. His forehead was pale, and he avoided looking at Ichiko and Takako. He walked into the living room, still in his overcoat, and rummaged through a drawer in the table near the window.

"Dear, Takako has come over. I wanted to talk with her, and I thought we might have lunch together. Will you be . . . ?"

"Oh." Chiba glanced at his watch as he turned. He was holding a folder of papers, which he carried with him to the center of the room.

"Sorry I was so rude," he said to Takako. "I'm afraid something rather awful has happened. I took a girl from work along with me to a construction site, and a carpenter overhead dropped a chisel—her cheek was cut."

"Oh dear. Was it Miss Mizuta?" asked Ichiko. Chiba nodded. Then he sat down in a chair at the table on which the bottle stood, the chair in which Takako had been sitting.

"Was she badly cut?"

"Yeah. I've left her at the hospital."

Ichiko turned to face Takako.

"She helps Chiba with his work. She's got a degree in architecture, of course. Chiba got her the job—she's very pretty. . . . It must be just terrible for her. I suppose she'll have a scar," she said. Then she noticed the color of Takako's face, and was surprised.

5.

Noticing as he washed his face that a cold drizzle was falling, Hirata—who had been sick with a cold for two or three days and was suffering from a fever—said he would stay home from work that day, and went back to bed.

"Use the phone next door and let them know at work," he told Takako. "Then you can come back to bed and get some sleep if you want to."

"I'm not really tired."

Stepping through the Chibas' gate, Takako saw that more sasanqua petals had fallen, forming a line alongside the hedge—they were wet. It looked as though there were more than just one or two days' worth—and the Chibas even had a maid. Takako wondered why Ichiko didn't ask her to sweep them up.

Ichiko seemed to be leaving—she came out wearing a raincoat, with her daughter in her arms. She waited in the entryway while Takako used the phone.

"Is he sick? Shall I drop by and say hello? I was just on my way to the hospital, actually. . . . I'm on my way to see the young woman whose face was cut by the carpenter's chisel."

"You're taking Yōko with you?"

"No." Ichiko put the child down and walked into the living room. She came back with three or four foreign architecture magazines.

"Chiba asked me to bring some of these to the hospital. They're doing plastic surgery—they say there'll be almost no scar at all."

The two women walked out to the street together. Takako stood on the doorstep for a moment, watching Ichiko's back as she moved off. Her flashy umbrella bobbed vigorously, keeping time with her rapid steps.

Takako felt a flicker of pain every time she realized that Ichiko was neither insecure nor jealous.

Chiba had looked very upset when he got home on the day Mizuta, the young woman, had been injured. It was perfectly natural for him to be excited and disturbed, of course—the person who helped him at work had cut her face, and he had taken her to the hospital himself. But Takako knew that he was feeling more than that. She had seen Chiba's love for the girl, exposed. She had come close to trembling.

The thought that Ichiko had seen her like that left Takako feeling weak. Ichiko, Chiba's own wife, wasn't at all jealous. Yet she herself, the neighbor's wife, was.

Takako had realized then that Chiba's emotions lay with Mizuta, that he only lived with Ichiko.

But if that was the case, why had Chiba returned home? Takako didn't believe that he had come to get the papers—none of them had seemed to be things he needed immediately. No, Chiba, unsettled, had come to see his wife's face. Part of him had needed to come home, if only for a short time. Takako was sure of this.

So Chiba might love Mizuta, but there was no question that he and Ichiko were man and wife.

Ichiko's umbrella bobbed out onto a larger street and was hidden behind a row of trees with yellow leaves. Suddenly Takako felt as though she was going to cry, and went inside.

"Takako!—Takako!" Hirata called, in a voice made nasal by his cold.

"Yes. I'm just going to wash some dishes."

Takako had a feeling she might break the china so she washed it with care, taking her time. She still hadn't finished when Hirata came in, still in his pajamas.

"What's wrong? You shouldn't be out here dressed like that."

Hirata put a hand on Takako's shoulder, playfully, then strengthened his grasp and embraced her from behind.

"Please—my hands are wet," Takako said, gasping. Hirata seized her wet hands in his fleshy palms and forced them roughly down against her skirt. She resisted lightly, trying to slip her shoulders out from between his arms.

"Are you and Chiba having an affair?" Hirata asked suddenly.

"What? Mr. Chiba? . . . Mr. Chiba and—why, there isn't—there isn't anything like that. There's nothing at all between us."

Everything grew dark as Takako spoke. Her knees buckled.

"You're lying. That's a lie."

He hit her in the face again and again. She fell to the floor.

"You think I don't notice that sort of stuff? You think I don't see the way your body changes?" Panting, Hirata put his hands around Takako's neck and shook her, violently. Takako opened her mouth. Her eyes were open.

"It's no good, I'm already . . . "

Takako spoke deliriously.

"No—it isn't no good. It isn't that it's no good."
Hirata answered her clearly, then lifted her up and
carried her. His arms felt stronger than usual—he
felt Takako's lightness. Hirata naturally forgave what
he could carry in his arms.

Takako wept cold tears. A second self watched the
self who lay there, not even trying to escape from
Hirata's arms. Her husband had mistaken the identity of her partner—he thought it was Chiba when it
was actually young Fujiki—but Takako knew that to
correct his mistake by telling him about Fujiki
would only enrage him again, only confirm that she
had been having an affair. Hirata's misunderstanding
about the identity of her partner left her a means of
emotional escape. He still didn't know for sure.

Takako grew numb as Hirata began to caress her.
She felt as if her affair with Fujiki had never happened. And—like the pain of a single pin piercing
her body—the pain she felt knowing that her husband thought she had loved Chiba flickered brilliantly. She found this strange.

"What are you thinking about?" Hirata shook her
roughly.

"Oh—you're cruel, you're cruel," she blurted.

Hirata fell asleep just as he was. Takako stared at
the chrysanthemums in the alcove. The flowers were
perfectly ordinary, but there were dozens of them in
the wide-mouthed vase, and in the dim light of the

alcove on that drizzly day they seemed imbued with the beauty of ghosts. Takako was reminded of the time a few days back when she had peered through the break in the bamboo fence at the Chibas' garden and seen the chrysanthemums and other flowers toppled messily over, but blooming.

Takako got up and walked into the sitting room. She rubbed her cheeks, thinking they must be red where Hirata had hit them. But it wasn't until after two that afternoon that she finally sat down in front of the mirror and put on her makeup, with even more care than usual. Hirata was still asleep. Takako had gone to tell him that lunch was ready but couldn't bring herself to wake him. When she had finished her makeup she began to wonder if he might not be dead, and looked back into the bedroom. But she still didn't try to wake him.

What would happen the next time her husband woke? Takako could hardly bear to stay in the house—she felt very weak, as if she was about to become terribly sick.

Ricky started barking next door, frightening her—it told her that someone was standing outside the fence. She walked outside on legs that shook. Of course Fujiki was standing there.

Takako's body stiffened, and she waved her hands at him. "Please—go away, go away! Why did you come here? What business do you have coming here? Didn't we agree not to meet any more? Please—please go away."

"I know. I know we've broken up, but—I just
wanted to see you once more. . . . Please—just
come this far."

"No, it's too dangerous. I'm scared. I can't."

"I'll leave right away."

"It's no good. I'm sick."

"Sick?"

"Don't just stand there—go away, please. This is
my house."

"Aren't you alone?"

"He'll be home soon, it's evening. And there are
the neighbors."

"That's why I'm telling you to come over here."

"Don't talk so loudly. . . . You'll just make things
worse. I'm sorry."

"I'll go. I'm going, so—please. Just come and say
good-bye to me."

Takako lifted her hand from the fence and hurried to the other side of the street, as if she were
escaping from something. Fujiki ran after her and
sheltered her with his umbrella.

"How far do I have to go? You'll kill me if you
come here again, you know."

Takako stopped at the corner of the tree-lined
avenue, planning to part with him there. Then she
cried out, "Oh!"

"I can't—oh, let me go, please let me go. It's our
neighbor."

Takako broke away from Fujiki and ran straight
toward Chiba.

Chiba was startled at the paleness of Takako's face.

"Why Takako, what's wrong?" He stopped and let her get in under his umbrella.

"I'm so sorry—that man . . ." Takako murmured, nearly collapsing onto him.

"What did he do?" Chiba looked in the direction Fujiki had gone.

"No, it was me. . . ."

"Well, he's gone. Shall we go home?"

"It's me—I'm no good. I'm no good."

Takako made no move to start walking. Chiba looked down at her, his eyebrows pinched together, puzzled. She felt an impulse welling up within her—not to confess, but to make an appeal.

"I—I had an affair with that man," she said. Suddenly her chest grew light, as if she had spit out poison. She felt a sense of freedom from herself that was at the same time a sense of her dependence on Chiba. Looking at her, he saw that the blood had drained from her face.

"I'll tell you all about it."

"Let's walk a little," said Chiba, leading Takako away from both their homes. She nodded. She was filled with a mysterious joy, as if she had already confessed everything to him.

But she didn't know how to start.

"Why did you? . . ."

"Why did it happen, you mean?"

"Yes."

"Because I'm a bad woman. I think it must have happened because I've been in love with another man for a long time."

"By another man, you mean the man I saw earlier?"

"No, I mean another man. . . ." Takako whispered, her chest trembling. "Hirata doesn't trust me—but the man he suspects is another man."

She heard herself speaking like a shameful woman plotting, yet the words made her happy.

"Does Mr. Hirata know?" asked Chiba, seeming genuinely concerned.

"Women are frightening—I really do believe that. I regret the affair so much I could die, and I've broken up with him, but once you make a mistake of that kind it's like a second self forms within you, and then you start liking the other person even more. It's frightening."

Chiba looked down at Takako again, a quizzical expression on his face. His blinked his black eyes two or three times.

"I'm really no good, am I? I only look obedient," Takako said. Then for a time she looked down at her own shoes and at Chiba's, which moved across the wet leaves as she walked.

A
Row
of
Trees

A row of giant ginkgo trees lined one side of the path that climbed the hill. Halfway up a narrow stone stairway led off to one side, down past a row of houses. The Soeda family's was the third in the row.

Soeda returned home from work at dusk on November thirtieth, and seeing the faces of his wife and daughter in the entryway, immediately asked them, "Have you noticed that half the ginkgo trees are bare?"

By "ginkgo trees" it was clear that he meant the row of trees on the path, but of course the words he had used conveyed little else, so he continued.

"I noticed on my way to work this morning, and let me tell you it gave me a shock. The ginkgos from the foot of the path all the way up to the area around our house are completely bare. But the trees from the middle up are still full of leaves."

"Really? I hadn't even noticed," said his daughter.

His wife's eyes said, "Is that so?"

"I wonder how it happened. Only the bottom half of the row is bare."

"And to think we haven't noticed. Shall we walk up and take a look?" his daughter asked her mother.

"It's dark already. Besides, you can see them from the second floor."

"Yes, that's true isn't it." His wife nodded. "You'd think we'd see them every day from upstairs. I can't believe we haven't noticed. . . ."

"Yeah, exactly. You ought to see them but you don't."

Soeda changed from his street clothes into the more comfortable clothes he wore around the house. It occurred to him as he did that the emotion he had felt that morning—the sense that he had made a discovery—wasn't quite as infectious as he had thought. His wife didn't seem at all excited.

After walking down the path that morning Soeda had looked casually back over his shoulder and then stopped, astonished. The ginkgo trees along the bottom of the path were completely stripped of their leaves, all the way up to their highest branches. And yet the crowd of trees up at the top was a thick mass of yellow foliage. It took about two minutes to walk down the path—it was short enough that the entire length of it could be taken in at a glance—and the row of trees wasn't long. So the split between the leafless trees and the yellow foliage had burst all at once upon Soeda's eye, leaving him peculiarly impressed. The bare branches of the enormous trees at the bottom of the path looked particularly sharp against the background of yellow leaves at the top, and the foliage that rose up over the path at the top looked richly colored and even more thickly layered than usual with all those bare trees in the foreground. The sense of great height that one feels looking at ginkgos was especially marked in these trees. Even the copious small branches of the bare trees strained toward the sky, as if attempting to embrace the tree-trunks, forming shapes tightly closed. The masses of yellow leaves conveyed a sense of volume as only layers of thick leaves can, but bathing in the morning sun they looked still and lonely.

The crowd of bare trees and the crowd of yellow-leaved trees didn't split neatly at a particular tree on the path, but generally speaking the division occurred about halfway up. Why it should occur halfway up—this too was a mystery to Soeda.

He passed under this row of ginkgos every day on his way to and from work. Several days had passed since he first sensed that the autumn leaves were starting to fall. But when had the trees along the bottom half of the path gone bare? Soeda had been so surprised by what he saw that he'd considered walking back up to the house to tell his wife and daughter of the change—this peculiar change in the row of trees.

He told them that evening. Of course neither of them had noticed.

"It's just as father says. I could see from the second floor," his daughter said, coming down the stairs.

"I thought you'd be able to. And you could see the split?"

"It's starting to get dark, but I could see it. Maybe I'll go up to the path and take a look." She walked out into the entryway. His wife was drinking some of the tea she had poured for Soeda, and made no move to get up.

"You're not going, Ikuko? Oh well, I guess you can go tomorrow morning. Though it wouldn't surprise me if the trees at the top lost all their leaves tonight."

"There's no wind."

"Wasn't it pretty windy a few days ago?"

It had drizzled, but there had been no wind.

The path rose, roughly speaking, from east to west. Thus, if a powerful easterly wind were to arrive, wasn't it possible that only the ginkgos along the bottom half of the path would lose their leaves? Soeda turned this thought over in his mind, but it seemed unlikely. That a row of trees planted, as far as one could tell, at the same time, long, long ago, should have split halfway up into bare and yellow-leaved halves—it was a natural phenomenon that Soeda lacked the knowledge to explain. And so he thought aloud, testing various ideas on his wife. The path rises from east to west, so the trees in the row must get approximately the same amount of sunlight on average, except that the way the light strikes the trees in the morning and the evening might be subtly different in the lower east and the upper west. . . . Or perhaps despite the fact that the wind hasn't been particularly strong lately something about the east and west winds has. . . . The ideas were all far-fetched, but still Soeda went so far as to close his eyes and concentrate on the topography of the area around the path. He knew the land so well that he could call its features up even without closing his eyes. But he couldn't see how those features related to the falling of the leaves.

"Well, at any rate, gingko leaves must be very sensitive to changes in something. I'm sure they just fall very easily."

Hearing her husband talk in this way, Ikuko concluded that he must be in a good mood. So she brought up a topic of her own.

"Today—well, it's rather an unpleasant thing to have to talk about, but something happened today that made me see what a truly good person Yūko is. It's incredible, really. I don't know, maybe she's becoming an even better person now that she's about to get married. That could be it. At any rate, I don't want you getting angry at her when she comes back, all right?" Ikuko's words anticipated Soeda's response, and checked it.

It had happened that afternoon while Ikuko was out shopping with the maid, at a time when their daughter was home alone. Yūko had carried a chair out to a sunny spot in the hall and was knitting a cardigan.

"Miss—I've got good soap and cosmetics out there, good wool too. Will you buy something?" Yūko heard a woman's voice and was surprised to see that the woman was standing quite close by. A hedge of full-blown sasanqua led in from the gate, and one had to pass through a wicket by the front door to enter the garden. But still the woman had come in—she was suddenly inside. Had the wicket been open? Had the woman opened it herself because it was low? Then Yūko saw that the young woman was carrying a baby on her back, and she relaxed her guard. The woman's sunburned face looked slightly swollen, but her hair was neatly arranged. She was small and plump, she had coated her lips thickly with a restrained shade of lipstick, and a weak smile hovered on her round face. She carried a somewhat large piece of cloth which she had folded to form a bag. There was in her none of the intimidating grimness of a peddler-by-force, but Yūko felt unsettled even so, and answered oddly.

"I have wool here, too. I don't need any more."

"I bet the wool I've got is better than that."

The woman came across the stepping stones to the stone that members of the family stood on when they removed their shoes, and standing there she peered rudely at Yūko's wool and squeezed it in her hand. Then, without commenting on the wool, she turned to face the garden.

"You've really got a nice garden, don't you? I wish I were rich enough to settle down in a house like this."

"I can't buy anything, but you're welcome to put the baby down and rest a little if you'd like."

"You don't mind?" The woman set her bag down in the hall and, without hesitating at all, took the baby from her back.

"You had everything so neatly arranged—perhaps I shouldn't have said anything," Yūko said.

She could smell the diaper.

"It can be real hard finding a place to nurse a baby when you're trudging around like this all day."

"She's adorable. How old is she?" Yūko looked down at the baby from her chair.

"Eleven months. People talk about how 'A child of one's own is never a burden,' right? Well, when you're carrying a kid around on your back all day it gets to be pretty heavy luggage."

The young woman lifted the bottom of her sweater, pushed her undershirt to one side, and placed her breast in the baby's mouth. Her breast was extremely full and faintly tinged with blue. The milk seemed to be coming out quickly, for the baby sometimes choked. White milk dribbled down from the corner of its mouth. Yūko drew closer to it and wiped its lips with her finger. She was enchanted by the way the baby's throat moved each time it sucked—she found it adorable—and thus thought nothing of the woman's full breast, which spread enormously before her eyes. The woman, too, was completely unembarrassed.

"Would you mind if I change her diaper here, too?" the woman asked. "There aren't many houses with people as nice as you."

Yūko watched what the woman did, and when everything was finished she picked the baby up. Her fingers and hands felt love when she touched the baby's skin, and for a while she was unable to let the baby go.

"You don't have any little ones here, do you?" said the woman.

"No, we don't."

"Are you an only child?"

"I have an older brother."

"You've really got a nice life here, don't you? Even someone like me—even I feel good here."

Yūko considered asking about the baby's father, but decided it would be a bad idea.

The woman crossed the stepping stones to the hedge and walked about looking at it. She seemed to be smelling the sasanqua.

"There sure are a lot of flowers, aren't there? They're beautiful."

Yūko wondered what the woman was feeling, what she was thinking as she looked at the sasanqua. Looking at the short, plump woman's back made her feel lonely.

Still holding the baby in her arms, Yūko went into the sitting room and brought back a purse. It was one she and her mother shared—the one they called the "kitchen purse."

"What kind of wool do you have?"

"Oh Miss, just letting me rest here was enough," said the woman. But even as she said this she was casually undoing her bundle, spreading it out. There were only two bundles of wool, one blue and one pale pink. Yūko bought the pink one.

The baby had been crawling about in the hall all this time, raising its voice in unintelligible cries.

"She's having fun. It feels good to be set free in such a big place."

Yūko asked whether the baby was able to eat biscuits yet, then stood and went back into the house to get some. She was only gone for a moment, but when she returned the woman had the baby on her back—she was preparing to leave. She accepted the small paper-wrapped package of biscuits politely.

"Thank you, Miss. I go from house to house, you know, but I hardly ever see a gentle face." Her own face reddened slightly, and she hid it. "If I come across anything good, I'll bring it—I promise. I'll be back, Miss."

Yūko watched the woman go, then lay the wool she had just bought on her knees and stroked it, remembering the feel of the baby's skin. She turned her eyes to the hedge of sasanqua next. She saw it there every day, so she had grown used to it—it was as though she had never seen it in full bloom before. There were so many flowers it almost seemed strange. But all the same, Yūko wondered again, what had the woman been thinking? What had she felt as she walked over to the sasanqua, as she looked at it? The woman's clothing hadn't been in the best condition, but of course the pink wool that lay on Yūko's lap was brand new.

It wasn't until sometime later that Yūko realized she didn't know what had become of the purse. It wasn't anywhere in the hall. It occurred to her that she might have returned it to the chest of drawers in the sitting room when she went to get the biscuits, but it wasn't in any of the drawers when she looked. Nor had it fallen into the garden.

"Yūko doesn't believe it was stolen," Ikuko said, having told Soeda the story. "She says the baby must have picked it up when it was crawling about in the hall. She says the mother must have put the baby up on her back without noticing that it had the purse in its hand, and gone off while the baby was still holding it. And if that was the case, the purse must have dropped out of the baby's hand and fallen somewhere along the side of the road. It's unlikely that the baby would hold it for very long. She searched the top of the path and the bottom of the path both, apparently—the whole length."

Soeda understood from Ikuko's manner of speech that the purse had not been found.

"Someone would almost certainly have picked the purse up if the baby dropped it on the side of the road. That's what Yūko says."

"She doesn't suspect the woman?"

"I'm sure she did suspect her, but she didn't want to. She says she can't even imagine that a woman like her would do anything wrong—that she just wasn't that kind of woman. And there was no question in her mind that the woman would come to return the purse if she had put it in her bag without noticing when she was packing. She expected her to come running back at any moment—apparently she couldn't settle down until I came home. And when she saw that the woman wasn't coming back to return the purse—that's when she decided that the baby had carried it off and dropped it."

Soeda had been warned not to scold Yūko, so he didn't say anything too hastily. It was possible, after all, that everything had happened just as Yūko said— that it had been an innocent act of the baby's, rather than a theft. The idea that the baby had carried the purse off and then dropped it struck Soeda as being quite nicely conceived. His mood softened.

"How much money was in it?"

"She had bought the wool, so—I'd say maybe two-thousand and six- or seven-hundred yen."

Soeda remembered that he had once mistaken a five-thousand yen note for a thousand yen note when he was climbing out of a taxi in the dark, and handed it to the driver. It had happened soon after five-thousand yen notes came into circulation. The driver had given him the change appropriate for a thousand yen note, and of course there had been no reason for Soeda to be at all suspicious of the man, no reason for him to look him over to see whether or not he had known that it was a five-thousand yen note. Soeda was inclined to think that at the time the driver hadn't noticed either, just as he himself had not.

He had told Ikuko about the five-thousand yen note at the time, but he didn't mention it now.

"Yūko hasn't ever had anything stolen from her before, has she?"

"When you say stolen—you mean, has she ever had anything of hers taken by someone?" Ikuko rephrased the question. "I wonder . . . something of Yūko's. . . . No, I can't think of anything. Maybe she hasn't."

They could tell by Yūko's hurried footsteps that she had returned.

"I got a good look at them," she said, coming into the sitting room. "It's not quite as exact as father said, but it is odd."

"What isn't exact?"

"The bare trees and the trees with leaves on them aren't really so clearly divided—they don't just split halfway up the path. There are still a few leaves on some of the trees at the bottom, and there are some at the top that have lost most of theirs."

"You checked every tree."

"Yes. The moon is out, and I could even see a few stars." Yūko looked Soeda straight in the face. "Father, have you heard about the purse?"

"I have."

"I'm sorry." Yūko's apology left Soeda unable to speak for a moment, and Yūko continued. "I've walked the path twice today. I was looking for the purse this afternoon, so I was looking down the whole time. This evening I was looking up the whole time, and I even saw the moon."

Soeda chuckled.

"I remember thinking this afternoon that a lot of leaves had fallen, but somehow I never noticed that the branches over my head were bare."

"I wonder if this sort of thing has happened before? Do you think the trees at the bottom of the path always lose their leaves first?" said Soeda, but Ikuko's only reply was to sigh and cock her head.

They had lived for years right next to the row of ginkgo trees, but none of them could remember if it was like this every autumn.

"There must be something wrong with us," Soeda muttered.

"We'll have to make sure to pay more attention next year," Ikuko said. Then, remembering that her daughter would no longer be at home when the autumn of that next year came, she was overcome with loneliness. "Why don't we send a letter to Shin'ichi in Kyoto? He hikes and he's interested in plants and so on—I wouldn't be surprised if he'd noticed."

"Maybe I'll take some pictures of the trees tomorrow and send them to him," Yūko said.

The next morning Ikuko walked with Soeda down to the bottom of the path so she could get a better look at the row of trees. Yūko came out after them. Every so often she would run on ahead of them, stopping to take a picture of the row of trees and of her parents standing before it. The whole thing turned into quite an event.

Three days later, late at night, a fierce winter wind blew up. December had begun. Soeda and Ikuko lay in bed listening to the sounds of the wind, discussing the ginkgos. The trees at the top of the path would be mostly bare in the morning, they agreed.

"The garden will be covered with leaves again," Ikuko said.

"Yes. I have to sweep them up, I remember. It happens every year."

The sound of trees thrashing in the wind came from the ginkgos in the row, there was no question of that. It seemed there might be another sound as well—the faint sound of ginkgo leaves dropping down onto the roof.

"It's lucky Yūko took the photographs when she did, isn't it? We can show them to Shin'ichi when he comes home for winter vacation. He says he hadn't noticed the split either."

Soeda understood that the sound of the wind had reminded Ikuko of their son. Shin'ichi's reply to her letter had arrived that morning. He had written that he didn't remember much about the stages in which the ginkgo leaves fell.

The half-leaved, half-bare row of ginkgos had seemed to Soeda like his own discovery. He felt now that the very last of its yellow leaves were being scattered by the fierce night wind—he could feel it in the chill at the back of his neck. Just as Ikuko had said, they would have to explain to Shin'ichi with photographs.

Shin'ichi had gone off to a university in Kyoto against the wishes of his family. Soeda still couldn't understand why he had been so opposed to the idea of attending one the numerous universities in Tokyo. The boy said that he liked the old Japan of Kyoto and Nara and insisted that the only time in his life when he would be able to see these cities as he liked was during his college days.

Soeda began to wonder, yet again, pointlessly, in the midst of the wind, whether Shin'ichi hadn't just wanted to try living away from home. And then one of Ikuko's little quirks drifted up into his mind. In autumn, when the fruits of the season started appearing in large quantities in the stores, she would buy them according to her color preferences. She liked the color of red apples, for instance, but disliked the color of tangerines. She ate tangerines, and in seasons when they were scarce—when she only saw a few at a time—these preferences seemed not to come into play. But she felt slightly uneasy when cucumbers and other vegetables were heaped up in the stores and often didn't buy them. Then too, there were times when she revealed an unexpected fastidiousness. It had happened more than fifteen years ago, but Soeda still couldn't forget the occasion when Ikuko had discovered a single scrap of toenail that had fallen onto the tatami—this despite the fact that he had cut his nails over a sheet of newspaper—and had picked it up, disgusted.

Soeda had gotten angry too. "I really can't understand it!" he'd said. "A piece of a person's body gets separated from his body, and all of a sudden you think it's dirty! You don't notice when you're kissing, but if someone told you to exchange saliva with someone else, you'd find the idea disgusting even if that someone else was a lover, wouldn't you?" The example had been rather unfortunate, admittedly, but Ikuko had been unable to forget those words for almost two months, and Soeda hadn't known what to do.

Perhaps that aspect of Ikuko's character had been passed on to their son, Shin'ichi? Shin'ichi was stubborn, it was true, but Soeda didn't think he was afraid of heaped-up cucumbers. And there was certainly nothing like that in Yūko. Suddenly an image of Yūko rose up in Soeda's mind, quite unrelated to what he had been thinking. It was Yūko as a junior high school student. She and a friend had been painting pictures, but now they had stopped, and they were coating each other's fingernails with red paint. Soeda tried to sketch in the details of this picture in his mind, to focus on it, and was soon oblivious even to the sound of the wind.

"What are you thinking about?" said Ikuko. "You can't sleep either?"

"What were you thinking about?"

"About the landlady at Shin'ichi's place in Kyoto."

Soeda had heard the story from Ikuko last year when she returned from a trip to see Shin'ichi and to sightsee in Kyoto.

"They went to put her grandfather's bones into the family grave when she was seven, and for some reason her mother said to her—You'll marry into a different family, you know, so you won't be allowed into this grave. She was just a child then, and she felt terribly lonely—but she was saying that the way things have turned out it looks like she'll end up in that grave after all! Can you believe it? She couldn't help laughing." Soeda could remember Ikuko saying this.

The woman was ten years younger than Ikuko. She and her husband had been childless when he

was killed in the war, so she returned to her parents'
house for six or seven years, then remarried into a
family with three children. She was fond of children, and the two young boys had grown to like her
very quickly—so much so that they fought to be
allowed to sleep beside her. But their older sister was
almost eleven, and she had been difficult. Once, at
her husband's suggestion, the woman had tried to
open an old chest of drawers in a room they didn't
ordinarily use, and the girl had struck fiercely at her
waist from behind.

"Stop it! Stop it! That's my mom's! She said she'd
give it all to me! You aren't allowed to touch it!" The
girl seemed on the verge of tears. The stepmother
tried her best to get along with the girl, but in the
end she gave up and left. Now she rented a five-
room house in Kyoto, the rooms of which she sublet
to students.

The fact that Ikuko had been thinking of the
landlady as she lay there unable to sleep suggested
something to Soeda—that her thoughts of their son
had led her to think of her own and her daughter's
lives as women. Still thinking this, he said, "It might
not be windy like this in Kyoto tonight, you know."

"That's true," Ikuko responded. Then, as if she too
was feeling better, "I suppose we'll go out again
tomorrow morning, all three of us, to see what's
happened to the leaves?"

"I'm sure they'll all have fallen."

In the morning the three of them walked together to the path, just as Ikuko had said, and looked at the ginkgo trees. A single night of winter wind had left them looking miserable. There were still a few leaves on the trees at the top of the path, but they were so thinly scattered that they ended up giving the trees a wintry appearance, making them look cold. Here and there among those trees that had still been leaved were ones that had lost their leaves entirely—that were stripped completely bare. The mysterious division that Soeda had discovered had collapsed. The row of bare trees at the bottom had only seemed splendid because the line of yellow-leaved trees at the top had served as a background. Even on the bare trees at the bottom a few leaves remained, scattered here and there, few enough to count. Soeda noticed that those yellow leaves were trembling, as if butterflies had come to rest on the branches.

Nature

To begin by saying that I heard the life story of a traveling actor at a spa is rather an old-fashioned narrative technique, but then . . . perhaps the story itself is old-fashioned. . . .

In June of this year, while traveling in Yamagata, it occurred to me that I might visit a certain spa—I had remembered that a spa once frequented by a now-dead friend was on the coast of Yamagata prefecture. My return to Tokyo would be delayed by a day, but since I was in the area anyway it seemed like a good time to stop by.

My friend had told me that the dunes and the sunsets were beautiful at this spa, and as my car cleared a stand of pine trees and approached the shore I saw that it was indeed lined with dunes. The stand of pines and the fields I had just driven through grew in sandy soil, and undulated gently— perhaps they had originally been dunes themselves. Or perhaps sand from the beach had made its way back even that far, even as far as the stand of pines and the fields.

As soon as I had been shown to my room on the second floor, I stepped out into the hall and looked at the sea. It was a little too early for the sunset, and I was unable to discover anything beautiful in the dunes I had seen from the car. They were bleak. There were large numbers of evening primroses, but they weren't yet in bloom, and if there were crinums and other beach plants growing as well, their flowers were not visible. It must be that, depending on the season, the time of day, and the lighting, there are times when those dunes become beautiful. It must be that at times—regardless of whether or not flowers bloom in the sand—the color of the sand itself becomes beautiful. The color of the sand and the colors of the sky and the ocean must reflect off and into one another, delicately merging.

Thinking that my friend, who had stayed at this spa frequently for long periods of time, must sometimes have seen the dunes and sunsets when they were beautiful, and gazing out at the sea, I noticed that the horizon between the water and the sky was unusually dark. Compared with the horizon I was accustomed to seeing—on the ocean to the west and south of Tokyo—the horizon of this northern region was unquestionably odd. Judging by the green of the trees I had seen on my trip, it seemed that the seasons here lagged about a month behind Tokyo's, but even so it was June: the guests at the spa were wearing light cotton kimonos—the sea should not be wintry. Or perhaps seas in the north were like this? Not only the horizon, but the color of the ocean itself was oppressive. I was amazed that my friend had been able to bear it, looking out on an ocean like this from his room at the spa, and I began to feel lonely. It occurred to me that the relative proximity of the aurora borealis might make the sunset beautiful.

When I asked the maid at dinner, however, I was told that my departed friend had disliked rooms facing the sea, that he had always chosen one from which the ocean could not be seen. It struck me as odd that he would come to a spa on the shore and then choose a room with no view of the ocean—I was a little surprised at first, but then decided that it made perfect sense.

"If you like I can show you his room later on," said the maid. "There's an actor staying there now, but . . ."

"If there's someone there—no, I don't really need to see it."

"The man likes novels. As a matter of fact, he was delighted when I told him that Mr. Kishiyama used to stay in the room. I'm sure he wouldn't mind."

I had signed the spa's guest book, and the maid had recognized that I was a friend of Kishiyama's, a fellow novelist. It seemed this maid had served Kishiyama whenever he visited the spa. He had frequented the place more than ten years earlier, in the years before the war, yet the maid looked fairly young.

"I suppose Mr. Kishiyama's daughters must be grown up by now," she said.

"The elder daughter is married—she had a child last year. The younger daughter is off studying in the United States. I served as the go-between at the elder daughter's wedding reception."

"Really? Then his wife must be living on her own now—she must be lonely."

The maid left the room and returned with a photograph and a scrapbook of notes and sketches. The photograph was one of the Kishiyama family, taken at the spa to commemorate a visit. The daughters had bobbed hair, and the maid was in the picture, too. It was a small photograph, but even so one could make out the faces of the people in it.

"Everyone's so young, even Kishiyama. . . . The older daughter really was beautiful, wasn't she."

Kishiyama had written in the scrapbook, in his odd handwriting, "All things join the flow of water." There was a blank space to the left of this, so I wrote, "Solitary shadow in stillness hoping for snow," and then at the bottom added, "Transcribed by Kishiyama." The words were not Kishiyama's own, but came from a Chinese poem he had written out for me. I finished writing and then savored the words. At any rate, it looked now as though Kishiyama and I had come together, had written together. Kishiyama had died seven or eight years earlier, so the sight of our characters lined up there, side by side, came to seem somewhat eerie. Kishiyama had died mysteriously at an airfield down in Kyushu—a field that was being used at the time as a base for the Special Attack Corps, the Kamikaze.

"Well, what do you say? The actor is waiting in Mr. Kishiyama's room right now, extremely pleased at the prospect of meeting you. . . ." said the maid, urging me to visit.

"He's come with a traveling troupe?"

"Well, he was acting in a play at a town nearby, and then he and our *panpan* became friendly, so he left the company and stayed on here alone. Now the *panpan* is supporting him. He's a handsome man, after all. . . . Everyone turns around to look when he walks outside." Seeing that the ink had dried, the maid closed the scrapbook. "When I say *panpan* I mean the sort of person who used to be called a *shakufu*, of course—a certain kind of waitress."

I went downstairs with the maid. She stopped before one of the paper-paneled doors in the hall, and called out,

"Mr. Uryū, Mr. Uragami has come."

He came to the door and slid it open, and our faces met. For an instant I felt as though I were looking at a large white flower. As we seated ourselves, Uryū came to seem more like an artificial flower—one which might easily be mistaken for a natural flower. After greeting me politely, he said, "I was told that this used to be Mr. Kishiyama's room, so I asked to be allowed to stay here—to remember. I read Mr. Kishiyama's *Collected Works* with great pleasure when I was in high school. . . ."

I doubted that a traveling actor would have attended high school under the old system, but I said nothing of these doubts. I said instead, "But to come to an inn on the shore and stay in a room with no view of the ocean? . . ."

"Ah—I have a terrible case of *muscae volitantes.* . . . I can't bear to look at the ocean or the sky or anything like that. The entire sky seems to be filled with flying ash-colored dots, they're like mosquitoes, and—" Uryū broke off. He squinted, as if dazzled. The look in his eyes was like that of a young woman flirting without even meaning to. I was drawn irresistibly towards those eyes. Hadn't I once known a young woman somewhere, briefly, whose eyes were beautiful in just this way—hadn't I been left with a sense of her charm, which was somehow like sadness and regret? Uryū's eyes were exactly like the eyes of a woman. They seemed moist with grief, but were in fact blue-white and clear. It was as though there was a second pair of eyes deep within the first. The feeling that I was being watched by that other, deep-down pair of eyes was strangely unsettling.

"When I see all those ash-colored dots flying . . ." Uryū began again. "It seems that it's my own life up there—that it's the world of my own heart and mind. My own sin, my own disgrace has become an ash-colored dust and filled the sky. . . ."

"Ah." For a time I was unable to respond.

There was a pair of sunglasses with light peach-colored celluloid frames on top of Uryū's desk.

"I've got *muscae volitantes* too, actually—though mine is a pretty mild case," I said. "Who knows, Kishiyama may have had it himself."

"Mr. Kishiyama wrote in an essay that he saw something like soot or tiny bits of thread, didn't he? But it was dust in the lenses of his glasses, not something that flew about. It moved when he moved his eyes."

"Did he really?"

How old could Uryū be? The smoothness of his cheeks and neck made it seem as though he might still be a young man. Deciding that he couldn't yet have turned thirty, I turned my eyes to the garden.

"To think that Kishiyama chose to look at a garden like this, rather than at the sea. There's nothing to look at."

There was a second two-story guest house opposite from where we sat. The garden between the two buildings was small and dark. The few trees in it had been cut close to the ground, and the rocks were small. Azaleas are fairly common in Yamagata, both as garden trees and as bonsai—I had seen azalea blossoms everywhere on my trip—but there were none in the garden here.

"I don't think Mr. Kishiyama looked at the garden. He would keep his face turned toward it, I'm sure, but that was all. He liked this room precisely because he could stay here without having to look at anything."

If Uryū was going to start talking like this, I would be able to talk too. I looked at the traveling actor again, thinking better of him. A knit shirt could be seen at the chest of his thin kimono. It was a shirt with no opening at the front, the kind one puts on over one's head. Such shirts are fashionable these days, but I found it slightly irritating. Uryū adjusted his collar and turned to face the garden. I was startled to see how beautiful his long, thin neck was in profile. It would have been long even for a woman's neck, yet somehow it didn't seem to be unnaturally long or thin—it was naturally and beautifully shaped. Its curvature was exquisite.

"The suggestion that he liked it here because he didn't have to look at anything is interesting, but then from what I hear you're an actor—could it be that you only like to show? Isn't there anything that you like to look at?"

"Well—I do like looking at nature's green. At trees . . ."

"Trees?" I nodded. "Yes, I have a fondness for big trees myself."

"I like forests and woods."

"But the idea of not looking at anything—a person who did that would be rather like Bodhidharma, wouldn't he. Nine years facing the wall."

"Yes, though the elderly are like that too, of course. Old people who've grown weak with age, who are near death, a natural death . . ."

"I guess so."

"Among my relatives in the country, there's a man ninety-seven years old, the oldest man in the village. Even his son is over seventy now. I went to visit in May and whether the old man was sleeping all day or whether he was awake I couldn't say, but he spent the entire day in bed. He lives in a separate building apart from the old main house—he's been living there for fifteen years. After the war the landlord's farmland was taken away, so that now the landlord has to work even harder than his old tenants just to make ends meet, and though there's a woman who looks after the old man, she's so busy with her work that he doesn't get much attention. I was told that she sometimes forgets to open the storm doors at the second house. At best she opens two or three of them about half way, just enough to let the light in. But on really bad days she just leaves them closed. It may be warmer if the doors are left closed on snowy winter days, I don't know. The point is that the old man seems to have no idea whether she's opened them or not. He doesn't even seem to notice when she leaves them shut, even when she leaves him in the dark at noon. I think that might be taken as evidence that the old man isn't looking at anything."

"How strange. And it isn't that he can't see?"

"I don't know. There are no signs that he has cataracts, his pupils are dark and clear—it seems most likely that he can see. I suspect that his mind has just stopped looking, though his eyes can see," said Uryū. His own eyes seemed to be seeing a vision of the ninety-seven-year-old man.

"Clear eyes would seem to be in your blood."

"The old man's eyes are so clear I sometimes think there must be Ainu* blood in him. He's always been pale, but since he hasn't been out in the sun for years his white skin has turned even more bluish and transparent. The way his skin looked I almost thought it would feel cold if I touched it, and when I looked at his arms it seemed that the color of the blood flowing in them must be thin, almost clear."

His hair and beard were white, glittering with silver. The seventy-plus son still had some black hairs mixed in with the white, but the old man didn't— there wasn't a single black or even slightly brown hair left on his head. And in the midst of all that bluish-white skin and silver-white hair . . . "His eyes, with their large irises, glittered with a black light—it

---

*The Ainu were the original inhabitants of what is now Japan. Tr.

really made a strange impression on me," said Uryū. "Those black eyes could see everything, and yet they weren't looking at anything. Thinking that made me cry."

"Yes. I suppose living out one's allotted life like that is nature's way—is a happiness nature bestows on its own. When a human being lives to be a hundred I doubt he needs Bodhidharma's Zen meditation just to die a natural death."

"Ah—during the war, it's been almost ten years, but at the time the old man was very sharp in certain ways. I was—well, I had turned into a person who didn't exist in Japan, a person who had vanished from the world—my fortune was such that I didn't know when I would die, or when I would be killed by my country—and so I started to long for the countryside, and I disguised myself and went there, and under cover of darkness I approached the old man's house. I stifled the sound of my footsteps, and the storm door was closed, but from inside there came a loud voice—Who is it? Are you a ghost or a thief? Is that Momosuke?—Momosuke is my name. I held my breath, but then he began trying to wake the attendant—Hey, would you open the door, Momosuke's ghost seems to have come. . . . I was terrified—I ran away as fast as I could. The old man must have been able to see me, right? Momosuke's ghost seems to have come. . . . Hearing that was terrifying. I can't forget it."

"When you say 'a person who had vanished from the world'. . . . "

"I had become a woman," Uryū whispered. "The man Momosuke had vanished. . . ."

"Yes—a woman," I murmured automatically, feeling certain things about Uryū fall into place. But I said nothing further on that topic, choosing instead to steer our conversation temporarily in a different direction. "How is the old man related to you?"

"He's the head of the main family. I'm in another line."

But the split had occurred generations back, and the blood relationship was distant, Uryū said. Uryū Momosuke's father and grandfather had been the local rebels, the local runaways. Momosuke's grandfather went to Tokyo and became a public servant and then sold the residence in the country. The house was dismantled and brought to Tokyo, where it was reassembled as a "country residence" in the garden of a certain house in Shiba. The house had survived the air raids without burning and was apparently still standing. Momosuke's father had become a military man. Momosuke, taking after his grandfather perhaps, rebelled against his father and ran away from home. His father, a commander in the artillery, resigned his position in the army because of this, but was immediately hired by a munitions company.

"Having a rebel for a son pained him so much that I hear he thought of committing *seppuku*, but now having lost the war it seems that his failing in the military on my account was really a good thing. He's still doing something or other, I hear. . . ." Uryū said, then lifted his face slightly and continued. "My father went to visit the old man in the country this past March. He went out to the other house and made his greetings, but from what I hear the old man just kept sleeping. He hasn't been awake much lately, the attendant said, with an expression on her face like she knew everything, and then suddenly the old man woke up, so my father introduced himself several times in a loud voice. The old man said—What was that? Toranosuke Hashimoto? I've forgotten the past, all of it. Then he went back to sleep, and the younger old man and my father smiled wryly and went back to the main house. I don't ever see my father, but I heard the story from the younger old man when I went out to the country a month later. . . ."

Momosuke's father's name was Uryū as well, but because there were a good many Uryūs in the village they had come to be known as the *hashimoto no* Uryū, the Uryū at the foot of the bridge, after the location of their ancestral home. This was shortened, and now they were called Hashimoto. Their old village lay on a narrow inlet, and looked as though it had once been level with the hills around it, and then collapsed. The green of the hills and the color of the sea were both very deep. Seen from a distance, the village looked like the vividly colored harbor of a tiny landscape constructed on a tray. Half of the town's residents were farmers and half were fishermen, with the farmland at the top of the hills. No one in the Uryū family fished.

"I've forgotten the past, all of it," Momosuke said again, repeating the old man's words. "We'd be saved ourselves if we could say that. . . ."

"Yes. But it'll be another forty years until we reach his age—until we're able to use those words naturally ourselves. It's a long time," I said. "It's a long time, but then unlike Kishiyama, who's dead, we're still living—which is to say that we've been granted a certain amount of extra time. Though of course, we ourselves have no idea how many more years we'll live."

"What?"

For some reason Uryū's face clouded.

"I've heard that the old man kept saying—I'll die in March, I'll die in March. People in the house would act amused and say—But it still snows in March, granddad, and there are days when it's terribly cold. . . . Why not put your death off a month, couldn't you choose a time when the cherry trees are flowering? But the old man would make a sour face and say—No, I'll die in March. Then in March of this year he caught a terrible cold and suddenly lost his strength, he stopped making sense when he spoke, so everyone thought—He's old, perhaps his time has come, and it will be March after all, just like he said, and they began preparing themselves for the funeral. Even the people who had been taking care of him lost their energy. But then for some reason the old man came back to life, as the doctor said, utterly amazed at it himself. My father and I went to the country to visit because of that illness. . . . But from the time of his sudden decline in March, he stopped saying that he would die in that month. . . . He doesn't say that he'll die anymore. He's completely forgotten about death. Or maybe that's not right—maybe he just goes on dying naturally, since from what I hear he spends all of his time asleep. He does eat quite a lot. But he just moves his mouth, he just chews—so everyone figured that really good sweets and so on would be like poison to him, that quantity should be more important than quality, but if they gave him cheap biscuits or something like that he'd grumble when he woke—Can't you people give me something better? He's able to taste things. In villages in the country they say that an old person who's become senile—whose arms and legs no longer work properly, who's always needing other people's help—they say that he's become a *nidoaka*. *Nidoaka* means someone who's turned into a baby for the second time."

"*Nidoaka*—that's a nice way of putting it."

"Momosuke's ghost seems to have come. Ever since the old man said that I've thought of him every time I'm in pain or in trouble, and I've grown fond of him—nostalgically fond. Mr. Kishiyama was an amazing man, but of course he died early. If he was alive and in this room now I think I'd want to ask him to live on to ninety-seven, to grow senile."

"Yes, but only one person in a thousand lives to be ninety-seven. If I live to be that old I think I'd like to be able to say, and to say naturally—I've forgotten the novels I wrote back then, all of them." I hadn't meant much by this, but Uryū was still a riddle to me, so I asked, "When you say that you had become a woman . . . do you mean that you were a specialist in female roles? Did you dress as a woman?"

Uryū lowered his beautiful eyes.

"I wasn't a specialist in female roles, I was an actress. I was living as an actual woman."

"Why? . . ." I asked foolishly. Imagining that something of Uryū's odd gender might be revealed in the area about his waist, I glanced towards it.

"Draft dodging. Hatred of the military. Fear of war," Uryū said coldly, quickly, slicing his speech into three. I was caught off guard.

"I was a high school student, but I didn't want to go on to university, so I became a woman."

"Ah. And you succeeded? . . ."

"Succeeded? I suppose you could say I succeeded. I made it through the war without becoming a soldier."

"Ah."

"These days, people talk about having been spies for the United States—reminiscences of that sort show up in magazines, even—and all the people who were prisoners then are walking about free, so I don't imagine that draft dodgers from back then will be brought up on charges—but I certainly made sacrifices in my flight, I certainly suffered. . . ."

"But you were able to live as a woman."

"I transformed myself into an actress touring the countryside. During the war people transformed themselves into all sorts of things—everyone's fate had gone crazy and so of course there were men who turned into women, either by themselves or with another man. There were probably a few besides myself."

"When you say that you became an actress, I take it you were in a company with other members. They didn't see through the disguise?"

"I wonder. . . . The head of the company certainly knew. I became friendly with the leader so that I could keep the rest from finding out, and the two of us were constantly together. The most difficult part was changing into costumes. I had to expose my chest then, of course. . . . I kept it wrapped up in gauze all the time, but even so. . . . I'm sure there were people who suspected, people who felt intuitively that something about me was a bit peculiar, but of course it never occurred to anyone that I had become a woman to evade the draft." Uryū smiled girlishly. "Even if there were other men who hated the idea of going to war—who hit on the idea of becoming women—not everyone can do it."

"Yes, that's true. If they aren't pretty like you. . . ."

"It isn't only a matter of being pretty or ugly— there has to be a woman inside one. . . . I think that there was a girl inside me. If there hadn't been a war that girl would probably have stayed clamped down inside, but we had a war and thanks to it she was able to push out into the world. I was told once after my transformation that there was a young woman who looked like me. All sorts of things happened to me during the period when I was a woman, but it's hard for me to talk about them so openly." Uryū stood and turned on the lights. The garden, too, had grown dusky.

"The sun must have gone down already," I said. "Kishiyama used to say that the sunset was beautiful here—have you seen it?"

"I don't look," Uryū answered abruptly. "Had you been planning to see the sunset? I'm afraid I've kept you. . . . I hope you'll see it tomorrow."

"No, I'm leaving tomorrow morning. . . ."

"So soon? . . ." Uryū said, moving his black eyes. "Would you mind listening to me talk about the time I spent as a woman?"

"No, I'd like to," I said, nodding. "I only came here because I knew that Kishiyama used to come."

Uryū nodded too.

"The *panpan* is supposed to come around eight-thirty, and I guess she'll leave around ten. Perhaps we could go for a walk on the beach afterward and I could tell you then—though I suppose you'll be leaving early tomorrow morning, won't you?" His inflection was feminine and somewhat pouty. I thought at first that he was embarrassed at mentioning the waitress who supported him, but that didn't really seem to be the case.

"My transformation into a woman was something I had planned out—even in high school I let my hair grow down to my shoulders, though schools at the time were very particular about things like that, and I was ordered repeatedly to have it cut. Eventually I got fed up and ran away. For a while I pretended to be a tramp in Asakusa Park, but the inspections were bothersome there too, and they were strict about keeping the number of drifters down. I changed over to women's clothing. Then I went back to my hometown, to the harbor. Actually I was born in Tokyo, so even though I say 'hometown' there are only a few relatives there who know my face. Momosuke's ghost seems to have come. . . . Hearing that nearly ninety-year-old man's voice come from inside the storm doors that moonlit night, I thought that I had left the world, that my life had reached its end."

"Ah—and then you became an actress?"

"That's right. I had liked acting ever since I was in middle school, and I had always played women's parts—I had been in the drama studies group at my high school, too—so I figured I might as well try to be an actress. . . . I had practiced a lot of the women's parts in various books of plays. Small traveling troupes were short on players during the war, so it was easy to get taken on. But even in groups of wanderers like that, people had their names registered—male and female actors both. Inspectors and officers in charge of conscription came around sometimes. I was the only one who wasn't registered. Uryū Momosuke had disappeared, no one knew where he had gone. Traveling actors are like fallen leaves blown together into a pile—they don't think very much about one another's pasts, and more often than not the stories they tell one another about themselves are made up."

"You became a woman and then acted woman's parts—that makes it a play within a play."

"A play within a play? . . . At the time it didn't feel like that. I was cursing the world and it felt good. The young soldier who told me about the girl who looked like me—he was in the Special Attack Corps. Our company went around performing to cheer up the troops. We went to a base with an airfield in Kyushu that was being used by the Special Attack Corps. There was a stream on the border of a wheat field, and on the other side of that stream a hill covered with all sorts of trees. I had been walking on the riverbank when I met this young soldier, and he stopped and stood there looking back at me, so I stood completely still, and he came back, and I asked—Did you see our play yesterday? Yeah, it was good. Then he told me that I looked a lot like his

girl. Really? How nice. Shall I show you her picture? Won't she be angry with you? Nah. Say—can
you jump this river? Of course I can't jump it. I'm a
woman. All right then, I'll carry you across. Oh
no—really! It was sometime in May, at the beginning of twilight. He carried me across the stream on
his back, and we hid under those trees, in the shadow of the leaves. He wanted to show me the picture
of his girlfriend. No doubt he wanted to look at my
face for a while. I didn't think that the young
woman in his tiny little photograph looked like me
at all. But I didn't tell him that. He talked about various things, and then he put his arm around my
shoulder and drew me closer to him, and tried to sit
me on his lap. I figured he'd probably realize I was a
man if I sat on his lap, so I grabbed his shoulder
from the side."

"I see."

"Are you a virgin? That's what he asked me. For a
moment my heart pounded. I'm a man, of course,
and 'Are you a virgin?' isn't a question one asks
actors in traveling troupes—it's the sort of thing that
might have made me burst out laughing, but I was
so startled that my voice wouldn't come. I shook my
head slowly. I certainly couldn't nod and so I shook
my head—but how did he interpret that? I see—he
said, stroking my shoulder gently. Even afterward I
couldn't figure out which he thought I had meant.
Aren't you frightened by these terrible air raids—
every day at every airfield? Yes, I'm frightened.

When I said that tears came to my eyes and I threw myself down on his lap and cried. Maybe he thought I had shaken my head unwillingly, and so he found me pretty. He was just a student, pure and unsophisticated. He said he would be sent to the front within two or three days. It occurred to me that if I hadn't transformed myself into a woman the same would be true for me. He lifted me onto his back again and we crossed the stream, and—what do you suppose he gave me as a memento?"

"What did he give you?"

"Cyanide."

"Cyanide?"

"Yes. His girlfriend had been conscripted to work in a factory, and she had asked for cyanide—as preparation for the worst. Of course at the time that sort of thing was quite popular among young women working in the factories. Apparently she had given a portion of it to the soldier. As preparation for the worst. . . . But he was going to die anyway—that was definite—so he said he didn't need any drugs."

"I see."

"I remembered that student again when I went to visit the *nidoaka*, the old man. As you see, I've gone back to being a man again. . . . When I went to the separate house the door was only open partway, it was very dim. It was May but the *kotatsu* was still there next to the futon. The younger old man, who had led me to the room, chased a fly. The ninety-seven-year-old had pushed the quilt down to his stomach—his right hand lay outside of the futon, and he was sound asleep. His white hair and his white beard had both grown long—he would have looked like a mountain immortal or a Buddhist saint if his color had shown a little more evidence of suffering, if he had borne scars of spiritual troubles—but the old man was too natural, he looked like a child, innocent. But after looking for a while at the fingers of his right hand, I noticed that his fingernails had separated from the skin already, that his fingers were entirely drained of strength. The younger old man called out in a voice loud enough to wake the older—Father, Hashimoto Momosuke has come to visit, and the area around the old man's eyes brightened, he gave a low groan and looked in my direction, and I was so startled by the glint in his black eyes that I gasped. The old man put his right hand down under the quilt and with the strength in that arm raised the upper half of his body. His white crepe obi had drifted up toward his chest. Quickly, say your name—urged the younger old man, but I just stared at his father. And just as something that has drifted lightly up sinks lightly down again, the old man's body dropped down, his right hand came slowly out from under the quilt again, and he went back to sleep. That was it. That will probably be our last meeting."

"I see."

"I couldn't help crying, but then that's a natural farewell, isn't it? Not like when I said goodbye to that student. . . ."

Uryū placed his knees together as a young woman would, and drawing his eyebrows together stared at me with his black eyes.

"I wanted to go back to being natural again, so I left the troupe and stayed on here by myself. . . . That company was different from the one I had been an actress with during the war. The first troupe was broken up by the defeat. I had gone back to being a man, a specialist in female roles, but still . . ."

I began to worry that it was time for the waitress to come. Like a large white flower Uryū allowed his body to soften, and hung his head.

Raindrops

The four children playing twenty questions downstairs in the children's room could be heard clearly on the second floor.

Taking turns, each child became the announcer, wrote a word on a piece of paper, and turned the paper over. All the players were children, so the words were simple. If a child's guess was "Correct!" a circle was drawn next to his or her name.

Apparently the child with the most circles would win.

"This one's a mineral," said the announcer.

"Is it a liquid?"

"Yes, it's a liquid."

"Is it water?"

"Yes, it's water."

"Is the water making noise now?"

"Yes, it's making noise now."

"Does it sound like—drip drip drip?"

"Yes. You're good at this!"

"Is it raindrops? . . ."

"Yes, it's raindrops! Correct!"

They had guessed the word with only four questions.

"Shizu, wasn't that Fumio just now—the one who said raindrops? He really is quick," said Hidaka Toshiko on the second floor, speaking through the sliding paper-paneled walls to someone in the next room. There was a three-foot hallway between the two rooms, so she and the person she was talking to were separated by two layers of walls.

"They certainly pick awful words. Raindrops of all things. . . ." Numao Shizu replied.

"It was really clever of Fumio, though. He's still so little."

"Of course Kaku, the boy from next door, he did say 'drip drip drip.' He let Fumio get the answer. My little Fumio is still just a baby, so Kaku lets him win.

He must have known it was raindrops if he said 'drip drip drip.'"

"I don't know. . . . You can't tell whether it's rain or raindrops just from hearing that it's water, and that it's making noise right now."

"If rain makes noise it's raindrops."

"That's not true. The sound of rain and the sound of raindrops aren't the same."

"Shin'ichi was the announcer just now, right? Well, that's why it was raindrops. Shin'ichi always picks unpleasant words like that. Kaku knew it right away."

Shizu spoke so decisively that Toshiko fell silent. There was no point in arguing over something as trivial as the difference between raindrops and rain. And the noise of the rain falling outside the children's room was really quite incredible. Water streamed down through a hole in the gutter—it was hardly the gentle sound that words like "raindrops" and "drip drip drip" described.

Toshiko had meant to be friendly when she spoke, and she thought now with some irritation that Shizu certainly might have been friendlier herself. Shizu, for her part, assumed that Toshiko was being sarcastic, commenting wryly on the fact that she was unable to repair the gutter, or for that matter the various other parts of the house that needed fixing. Her fifth-grade son's choice of that wretched word, "raindrops," had put Shizu in a bad mood.

Toshiko and Hidaka were a young couple, so recently married that they hadn't yet been registered as a family. They rented the six-mat room on the second floor of the Numaos' house, and they both worked. The first floor of the Numaos' house was a stationery store. It had managed to remain in business because the elementary school was nearby, but Shizu had stopped bothering with inventory ever since Numao started staying away from home, and she'd developed a habit of speaking curtly to children when they came in to buy things, and sales were falling off as a result. Shizu had the impression that the young Mr. and Mrs. Hidaka were on good terms—that they were an example of that modern rarity, a friendly couple—and so she behaved toward them with a mixture of goodwill and envy. She was jealous on some occasions, and on some occasions she was downright mean—her moods varied. In the past Numao had slept by himself in the eight-mat room on the second floor, and Shizu had slept downstairs with the two boys, in the six-mat room at the back of the store. The children's room was an addition—a narrow room with a floor made of planks, into which the two boys' desks just barely fit. When Numao didn't return home for several nights in a row, Shizu started sleeping in the eight-mat room on the second floor. The children still slept downstairs. Numao's loud snores had always made the Hidakas feel more comfortable somehow, more secure. Shizu, on the other hand, often startled

Toshiko by speaking to her from the far side of the paper-paneled walls at times when Toshiko felt sure she must already have gone to sleep.

"Toshiko? Toshiko?" she once said. "You know, I've been thinking. I bet the woman Numao is seeing, I bet it's that Tokiko, the one who used to come talk with you in your room all the time. She hasn't been around at all lately, has she?"

Tokiko was a young woman, still unmarried, who worked at the same company as Hidaka and Toshiko. Toshiko suspected that she was the woman *her* husband was seeing, and that their affair was the reason Tokiko no longer came to visit her in the room she and he rented.

Toshiko had been crying and complaining, whispering tearfully to her husband when Shizu spoke that night. Hidaka seemed disturbed by Shizu's delusional nighttime mutterings—he used both his hands to push Toshiko's head from his chest. Toshiko lay still for some time.

"That's a frightening voice, isn't it?" said Hidaka.

"You can do whatever you like when I'm like Shizu—when I've got a child in fifth grade. I swear I won't care in the slightest. But now—now!—it's unbelievable! We haven't even been married a year, and already. . . . You realize, of course, that Miss Tokiko is having other affairs—she has to be, with Shizu saying things like that."

"Yeah, I know. You're a suspicious devil just like Mrs. Numao. Just remember that voice of hers."

After that night Shizu never said anything about Tokiko again. Tokiko sometimes dropped by to see Toshiko, and on those occasions she would talk pleasantly with Numao as well. Surely Shizu didn't suspect even then—in broad daylight—that Tokiko and her husband were up to something?

"Better watch it—she's in a dangerous mood," Hidaka warned Toshiko after the incident with the word "raindrops" in the game of twenty questions. The four children were still playing the same game. Fumio—the boy who had guessed the word raindrops—had to be given assistance when it was his turn to be the announcer, since he was only in second grade.

The siren of an ambulance approached through the darkened streets. It felt as though something awful was about to happen.

"Oh, I hate it! I've heard that siren three times today," Toshiko said, speaking loudly enough that her voice would be heard in the next room.

"Yeah, there are lots of accidents at the end of the year. People are in a rush, and cars go too fast—it's dangerous. I heard somewhere that when ambulances get sent out it's almost always for people who've been injured. They hardly ever go out for sick people. I'm sure the cars must be skidding like crazy in this rain," Hidaka said.

"That sound—it's like my chest is stopped up. It makes me feel a little old. Like a torrent of years is rushing at me."

"Here we are sitting nice and cozy in the *kotatsu*—why do you have to talk like that? Someone's gotten hurt."

Toshiko lowered her voice. "It's not funny, you know. I'm going to have to quit my job after New Year's, right? You said so yourself—the company has its rule, if two people in the company get married one of them has to quit. And who told everyone at work that we were married? It was Miss Tokiko, wasn't it?"

"It was nice being able to keep it secret for half a year."

"It wasn't the slightest bit nice. When I think that I'll have to quit my job next year, it's like there's an ambulance speeding through my chest."

The siren moved off through the streets.

A few moments passed, and then Shizu spoke again in the next room.

"Toshiko?" she called. "You know—I don't know why, but for some reason ambulances go by a lot on this street. It's been that way for some time. Every time I hear that siren I think how nice it would be if my husband were run over by a car—if he were killed, or even injured."

". . . . ."

"If he were injured he could still look after the store."

Hidaka and Toshiko were staring at one another. Toshiko was unable to move her adorable, still-youthful eyes from Hidaka's face.

"Toshiko? Have you gone to bed?" Shizu said.

"No, it's still early. It's only nine twenty," Hidaka answered in her place.

"Is it? Then do you mind if I come talk?"

"Not at all," Hidaka answered again.

Toshiko glanced up instinctively at the sliding door, and adjusted her position under the quilt of the *kotatsu*.

Shizu began speaking as soon as she had climbed in under the *kotatsu* with Hidaka and Toshiko, who were sitting face to face.

"Mr. Hidaka—I've been thinking I might rent out the eight-mat room next year. What do you think? If there's someone among your acquaintances you think might be good, would you let me know?"

"Of course."

"I won't have anyone like Miss Tokiko."

"She lives with her parents."

"What? With her parents?" For some reason Shizu seemed surprised. "Well isn't that nice for her. But at any rate, if I can rent the eight-mat room, I think I'll use either the key money or the deposit to fix the gutter. It makes such an awful racket every time it rains—I feel just terrible for the children. It's so loud I can't sleep myself when it rains hard."

"Surely it's not that bad?"

"It is that bad. I honestly can't sleep. But what I wanted to say is that if someone's going to be moving in next door anyway, I thought it might be nice if it was someone the two of you know. A young couple would be best, I think."

"It really doesn't have to be anyone we know—we don't mind, do we?" Hidaka asked Toshiko.

"I'd like to have two young couples on the second floor—two couples who really get along well together. Then I could try to make things work with the children in the small room downstairs. I'd like it to be a young couple. You chose the print for the quilt on this *kotatsu,* didn't you, Toshiko? Did you sew it yourself? It's nice."

The front door rattled open downstairs.

"Shin'ichi? Fumio? Are you here?" They heard Numao's gruff voice. "Hey, hey, there you are! You know, out there just now, boy was that terrible! It was so sad I couldn't stand to look. Someone's boy got hit by a car."

The four children had evidently stopped their game of twenty questions when they heard Numao's shout. It sounded as though they were going out to meet him.

"He's come back," Shizu said. She stood up hurriedly from the kotatsu, and then, as if it embarrassed her that Hidaka and Toshiko had seen her hurrying for her husband, said, "It must have been for the boy—the ambulance earlier." She left the room.

Suddenly Tokiko's voice came from downstairs. Shizu must have criticized Numao for coming home with Tokiko, judging by the way he now spoke.

"Me? No, no, let me tell you—this person right here! She's the cruel one. A little kid gets hit by a car, and what does she do? Just like a woman, she's there mixed up with the rest of the rabble, staring. I couldn't believe it. But what can you do? A woman with no children of her own is like that."

"Oh, you're terrible! I thought it was awful. Besides, I seem to recall that you were looking, too, Mr. Numao."

"I was. I heard it was a child, and I thought it might be one of ours. So I pushed my way through the crowd, right, and when I got out in front Miss Tokiko was standing right there!"

"What happened to the child? Will he live?" Shizu asked.

"How would I know? I saw that it wasn't one of our kids, and I just felt so relieved that I. . . . I saw them put him in the ambulance, but that was it."

"See what I mean? You see that it's not one of your own children and you feel relieved. You don't think that's terrible?" Tokiko said to Numao.

"I think you're absolutely right," Shizu said, siding with Tokiko. Then, changing the subject, "Whenever someone from our house is out and I hear an ambulance or something like that, I hate it."

Tokiko came up to the second floor. The room seemed to get brighter. When Tokiko still hadn't spoken after a few moments, Toshiko spoke.

"We heard the ambulance siren here, too," she said.

"Did you? That accident made me feel awful. I was on my way here to tell you—well, that I'm going to be getting married before the end of the

year. It's very sudden. . . . He's quite a bit older than
I am."

"Well, that's wonderful! I had no idea!" Toshiko
said brightly. For a moment it seemed that her pretty eyes might fill with tears. "Darling, if Tokiko
needs a room—how would it be if she took the
eight-mat room next door?"

Hidaka didn't respond. The lovely shadow of
Tokiko's nose moved on one of her cheeks, taking
on peculiar shapes.

Various sounds indicated that the children from
next door were going home, and then came the low
murmur of the couple talking downstairs. The
sound of the water pouring from the hole in the
gutter outside the children's room drowned out
their voices.

Chrysanthemum in the Rock

What sort of rock had it been? I looked through a number of books like Awazu Hidekōji and Wada Yaezō's *A Full Color Guide to the Rocks of Japan* in an effort to help myself remember, but I still couldn't be sure. I'm entirely ignorant as far as rocks and stones are concerned, and I'm sure it would have been difficult for me to decide which of the illustrations the rock matched even if I had spread the chart out in front of the rock and compared the two directly. To make matters even worse, it'd been thirty years since I'd seen the object itself. The rock is in my hometown, far away.

There was a large hollow in the front of the rock that someone had packed with soil, and a chrysanthemum had been planted there. I remember seeing the white flower in bloom, a chrysanthemum so swollen with petals it looked like a ball. Nowadays large flowers of this shape are sold in every florist in Kamakura. Maybe that ancient chrysanthemum was a flower of the same variety, maybe it had simply gotten smaller over the years? It grew on the surface of a rock, after all, and there was no one to look after it.

The blossoms of the white chrysanthemums that florists in Kamakura sell are extremely heavy—so heavy that if you stand one up in a narrow vase the arrangement looks unsteady, as though the vase is about to fall over. The chrysanthemum that blossomed in the rock was small, but serene.

And it hadn't been planted there on a whim, or to comfort the living. It had been planted as an offering.

A woman's head had been appearing over the rock. It was a ghost. A memorial service was held, a chrysanthemum was planted in the rock as an offering, and the woman's head stopped appearing. Ever since then the villagers have planted a chrysanthemum in the hollow in the rock every year. This was the story people told.

I haven't visited the town where I was born in thirty years, and though I see chrysanthemums every year I'd somehow managed not to think of that rock until this autumn, when it occurred to me for some reason that it might be considered a kind of memorial, a kind of gravestone.

I had been walking from temple to temple in Kamakura, looking at old stone art.

And I'd been telling people, "There's not much from the Kamakura Period left in Kamakura, but I'd guess that most of the pieces that do remain unchanged were executed in stone."

Consider a few examples. The treasure-box-seal stupa dedicated to the founder of Kakuen-ji, the treasure-box-seal stupa of the same temple's second head priest, the five-ring stupa marking Ninshō's grave at Gokuraku-ji, the stone torii* at Tsuruga-oka-hachimangu, and the seamless stupa at Daikakuzen-ji, which is one of the temples within Kenchō-ji—all of these were designated as National Treasures under the old system. The treasure stupa at Betsugan-ji and the treasure-box-seal stupa erected at Uesugi Kenpō's preemptive funeral have been classified as Important Works of Art, along with a number of other works, while the treasure-box-seal stupa dedicated to Hōjyō Shigetoki has been designated as a historic place. The kurikara monument at Gosho-jinja, the seven-layer stupa dedicated to

---

*A gateway found at the entrance to all Shinto shrines. Tr.

Hōjyo Dōgo, and the stone Buddha at Fukōmyō-ji also come to mind. The Kamakura Period was the great age of stone art.

And yet there aren't many people who walk about looking at artworks of this sort. I myself have been living in Kamakura for fifteen years, and until this autumn I never felt any curiosity to see them. No doubt this is because most of the pieces are old gravestones.

"After all, they're gravestones," I murmured to myself as I set out alone on my tour, without even asking my wife to join me.

But to tell the truth, the fact that so many of the pieces are gravestones is what prompted me to make a tour of them in the first place. Several of my friends and acquaintances had passed on, and as their graves were completed I had had the occasion to see quite a number of gravestones of various shapes. One stands before a grave and thinks about the deceased, and so naturally one starts thinking of the shape of the stone as well.

One of my friends had had a small treasure-box-seal stupa built for the grave of his wife, who had passed on ahead of him. He explained to me that the form of the treasure-box-seal stupa was based on that of the gold-lacquered stupas built by Qian Hong Shu. Following the ancient example of Emperor Ashoka, who had been a great builder of stupas, Qian Hong Shu, the king of Wu Yue, ordered his craftsmen to fashion eighty-four thousand stupas out of copper. He had a copy of the *Treasure-Box-Seal Shinju Sutra* placed in each stupa, and then sent them to various countries. Some of these stupas made their way to Japan. The small gold-lacquered stupas were constructed in the year that corresponds to the ninth year of Tenryaku in Japan. Stone treasure-box-seal stupas were first used to memorialize the dead sometime after the beginning of the Kamakura Period. It is generally agreed that truly beautiful works of the treasure-box-seal stupa form were produced exclusively during the Kamakura Period.

I've lived in the valley of Kakuen-ji, the two-storied temple, for more than ten years. I have sometimes visited it on my walks, there at the farthest end of our small valley, and any number of years have passed since I first saw the two famous stupas in its precincts. I didn't know until recently, however, that these stupas—one of them dedicated to the temple's founder, the other a priest's stupa erected for Master Daitō, the Sixth Patriarch—are considered the largest and most beautiful treasure-box-seal stupas in the entire Kantō region. The upper portion of the stupa dedicated to the temple's founder toppled off during the disastrous earthquake in Taishō★ twelve, and I've been told that if you looked at it then you could see two holes in the second of the stupa's four sections. One of these holes contained the priest's bones.

★The Taishō Period began in 1912 and ended in 1926. Tr.

The stupa dedicated to Tada Mitsunaka, generally known as his grave, which I've seen any number of times through the windows of the bus that crosses Jukkoku pass, down at the bottom of the mountain road—it too is a treasure-box-seal stupa of the Kamakura Period. Izumi Shikibu's stupa, also generally known as her grave, which one sees in Kyoto amidst the hustle and bustle of the Shinkyōgoku arcade—it is also a Kamakura Period treasure-box-seal stupa. The latter is twelve and a half feet tall, making it less tall than the stupa at Kakuen-ji, which rises to a height of just over fourteen feet. But the treasure-box-seal stupa's form is beautiful and delicate, and a small stupa of this type seems suitable for a woman's grave.

Though I live in the valley of Kakuen-ji, with its beautiful gravestones, the first time I considered a gravestone as an object of beauty was when I saw the treasure stupa at Sen-no-Rikyū's grave and the stone lantern at the grave of Hosokawa Sansai, both of which are found in Kyoto's Daitoku-ji. The stupa and the lantern were both stones that Rikyū and Sansai loved during their lifetimes—stones that they decided to use as gravestones—and so we look at them from the very first as objects that Rikyū and Sansai found beautiful. Because Rikyū and Sansai were masters of tea, because the stones are linked to the world of the tea ceremony, one feels an intimacy and a brightness in them that most old graves lack.

The place where the figure of the door would usually be carved on the lowest of the three segments of Rikyū's treasure stupa has been scooped out, and it's said that if you put your ear in the hollow you can hear the sound of water boiling for tea, a sound like wind whispering through a forest of pines. I put my own head in the hole. My thin face fit perfectly inside—so perfectly, in fact, that the bones of my cheeks rubbed slightly against the stone as I drew my head out.

"Could you hear a kettle?" someone asked, and I replied, "I suppose so. I guess if you listen for it there does seem to be a sound of some sort." Even as one puts one's face up into the gravestone, one thinks more of listening for Rikyū's kettle than one does of the fact that the stone is a grave.

"Legend has it that Rikyū liked this stupa so much he stole it from the grave of the retired Emperor Nijō, at the foot of Mount Funaoka," I said.

It's said that treasure stupas have their origins in the eleventh chapter of the *Lotus Sutra,* "The Appearance of the Treasure Stupa." When Shakyamuni preached the *Lotus Sutra* on Mount Gridhrakuta, a stupa adorned with the seven treasures erupted from the ground and hovered in the air. Out of that magnificent stupa there came a voice which praised Shakyamuni extravagantly. With the fingers of his right hand Shakyamuni opened the door of the stupa, and there sitting in the lion seat was the Buddha Many Treasures. Many Treasures offered Shakyamuni half of his seat. "Then the

assembled priests saw the two Buddhas sitting with
their legs crossed in the seat of the lion in the stupa
adorned with the seven treasures, and they all
thought, 'The Buddhas sit in a place high above us,
far away. Oh, how fervently I wish that the Buddhas
would use their godly powers to bring us up into
the sky with them!' And instantly the Buddha
Shakyamuni made use of his godly powers to sum-
mon the many priests, to carry them into the sky."
Then Shakyamuni said, "The Buddha Many Treasures is usually found wandering in the ten directions. It is only for the sake of this sutra that he sits
in this treasure stupa now." Thus wherever the *Lotus
Sutra* is preached, the stupa of the Buddha Many
Treasures appears.

It is for this reason that doors are carved on the
front, and sometimes on all four sides of the so-
called "axle" of the lowest of the three segments
that make up a stone treasure stupa. Rikyū's stone
stupa—scooped-out where its door ought to be,
and carved entirely from one rock, though it's more
than six feet tall—is unusual even in terms of its
shape.

There is a story to the effect that Sansai carried
his lantern with him even on the biannual journeys
to Edo that were required of all daimyo by the
Shogunate—journeys that he made on foot. This
may or may not be true, but whatever the case, it's
certain that Rikyū's and Sansai's gravestones were
not carved for them by stonemasons after their

deaths. They are products of an age even more ancient than that of Rikyū and Sansai, stones which the two viewed as works of art while they lived. And the beauties they saw in those stones while they lived became their graves just as they were. It's certainly an interesting way of making a grave. The aesthetic spirit of the person buried in the grave assumes the form of a gravestone.

Rikyū could probably have sketched out a design for precisely the kind of treasure stupa he wanted in his head, but it's unlikely that he would have ended up with an artwork as beautiful as the one he imagined if he had given the design to a stonemason and asked him to carve it. All the strengths of the age couldn't have helped. And of course, there's a certain quiet beauty that only certain ages produce—one sees this even in stones. It's true that stone lanterns of various unusual shapes started reappearing around the beginning of the Momoyama period—lanterns carved to suit the tastes of those involved in the tea ceremony—but even so it seems reasonable to say that the designs only grew less and less interesting following the end of the Kamakura period, that there was a general stylistic collapse. Rikyū and Sansai selected stones they liked from the works of art that older ages had left them—works that their own age lacked the power to produce—and they used these stone artworks as gravestones. Perhaps this was the extreme limit of extravagance, of arrogance. But then on the other hand perhaps one might think of it as a pure elegance, as humility. Even those of us who visit their graves now, generations later—who can say how much our sensitivity is honed and deepened by the ancient treasure stupa, the old stone lantern? One might say that Rikyū and Sansai became so obsessed with their stones—the one with his treasure stupa and the other with his lantern—that they carried them all the way to the graves they were buried in, but one might also say that they continued to the very end to value the beauty they had cultivated their entire lives, that they aestheticized even their gravestones.

Thus the first things that came to mind when it occurred to me to make a tour of Kamakura's stone art were the graves of Rikyū and Sansai.

I was looking at a collection of photographs of stone artwork one day, when my wife looked down over my arm at the book. Seeing that most of the photographs were of graves, she asked, "What kind of gravestone shall we have made for you?"

"I'm going to buy an old stone, something I like," I replied. "I'd rather choose one myself." Stone stupas and various other pieces that would make good graves are still bought and sold as old art. If I had to have a gravestone, I might as well learn from Rikyū and Sansai and choose a stone that pleases me while I'm still alive—thus in my eccentricity I daydreamed. A many-treasure stupa, a treasure-box-seal stupa, a five-ring stupa, a seamless stupa—perhaps even a stone Buddha or lantern would be nice. The very thought of having an old and beautiful stone marking my grave lessened immeasurably the pain I felt on considering what would, after all, be my burial. People who came to offer prayers at my grave would feel beauty. That beauty would be something I had seen, but not something that either I or my age had produced. It would be a beauty no longer capable of being produced. A beauty left to us by ancient Japan, a stone that would not decay, that would remain exactly as it was for ages to come. Seen in the stream of that stone's long life, my own life would seem quite short.

I would use the old stone artwork just as it was, I wouldn't even have my name and age carved into it. Only people who'd been told would know that it was my grave. But even those who didn't know would be struck by its quiet beauty as they passed. And even after the people who knew it as my grave had all passed on—even then my gravestone would stand there in its place, beautiful, preserving within itself one aspect of Japanese beauty.

There's really not much reason to think about the gravestone one will have when one is dead while one is still alive, but as the gravestones of more and more friends and acquaintances are erected over their plots, there come times when such thoughts flicker up in one's mind. Gravestones have been built even for people who insisted they didn't want one. I myself have always thought of a gravestone as a sort of heavy burden, which probably accounts for my dreams of having a stone like Rikyū's or Sansai's. It's unlikely that I would end up with anything beautiful if I were to have a gravestone carved for me after I die. This is one of the depressing facts of these latter days of the Buddhist law.

An acquaintance of mine who works as a dealer in old art told me that he had one thirteen-layer stupa from the Kamakura Period among his things, and so, looking out at our untended garden, at the grass I had let grow in a wild tangle, I wondered if it wouldn't be pleasant to go ahead and buy it, to have a thirteen-layer stupa out there, shooting freshly upward.

"Wouldn't it be great to have a Kamakura Period thirteen-layer stupa right in the middle of our garden? That's all I'd need."

I didn't mention that it would become my gravestone when I died.

The young art dealer said he could probably bring the pieces of the stupa to the house in a single truck. . . .

"But it will be hard to stack them up. It's a good twenty-one feet tall, you realize. We'll have to build something tall to stand on and then just carry the stones up."

Even assuming that a thirteen-layer stupa could be used as a gravestone, a stupa twenty-one feet tall would be visible from quite a distance—it was a bit too tall to stand over my grave. It would give people too much of a shock.

The seamless stupa dedicated to Rankei Dōryō, the founder of Kenchō-ji, and the one dedicated to Mugaku Sogen make beautiful gravestones. There is a profundity and an elegance in the egg-shaped bodies of these stupas, which are supposed to be formless, to contain all things in heaven and earth. When I see calligraphy by Rankei or by Mugaku at a tea ceremony, the shape of the seamless stupa—of their gravestones—sometimes drifts into my mind. There is a row of seamless stupas in Kakuen-ji, too, carved for successive generations of priests, their egg-shaped heads lined up. I like seamless stupas, I find that they make nice gravestones, but then it seems that all seamless stupas mark graves. If possible I'd prefer to avoid having an object that has been used as a gravestone already, for some other person, marking my grave. Of course, if I had a seamless stupa carved for me now, in this world, there's no doubt that it'd end up being ugly, having a form unpleasant to look at. Once again, the most beautiful seamless stupas were produced in the Kamakura Period.

Now that I think of it, I'm really quite lucky to be living in Kamakura, where I'm able to see all these works of stone art on my walks. Stone architecture didn't exist in ancient Japan, and no large works of art were executed in stone. It has been said that this is an indication of the fragility of Japanese culture, and it's certainly true that the seamless stupas, the treasure-box-seal stupas, the five-ring stupas, and the stone Buddhas I saw in the old temples of Kamaku-ra looked every bit as old as they were, that they had about them an air of loneliness and poverty. They were old stones of the sort one might find hidden away in the shadowed recesses of a mountain, possessing none of that beauty which causes one to stand gazing up in awe. And yet the longer I looked at those stones the more I was struck by a sense of the powerful beauty that filled them. I felt an indescribable affinity with Japan's past.

Then, walking home from my trip to see the seamless stupas, stepping through the fallen autumn leaves, I was suddenly reminded of the rock in the village where I grew up, and the chrysanthemum in the rock.

If one thinks of the rock as the grave of a woman with no grave, would the chrysanthemum then be an offering? The story concerned an unfortunate, nameless woman. It was the sort of story one hears all the time, of a woman from a family living in the mountains. The woman was waiting for a man behind the rock, and she froze to death. That's the story.

The village I come from lies in a valley with a river running through it, and there are many rocks around it, both lining the banks of the river and within the river itself. The rock behind which the woman waited was one of the largest of these—there was no danger that she would be discovered if she stood behind it, even if someone were to pass by.

There was a small pool at the base of the rock. The shadow of the rock was so big that it filled the pool, that the pool could not contain it. The woman must have been waiting very eagerly. From time to time she would climb up onto the rock, raise her head over the top and look down the road along which the man was supposed to come. The ghost that appeared, the woman's head floating over the rock, must represent the woman as she was at those times. The woman climbed up to the hollow in the rock and stood there. And so that's where the chrysanthemum was planted.

I went out the main gate of the old temple in Kamakura and, walking along under the row of cedars, spoke to the ghost of the woman whose head floated over the rock in my hometown.

"Your long hair is wet. Did your tears stream down into your hair? Or did tears stream even from your hair?"

"It must have gotten wet in the sleet yesterday. I'm happy waiting here for him. I have no reason to cry."

"It looks like it's going to snow again tonight. Go home quickly—you'll freeze if you don't. I doubt he'll come today, either."

"He told me to wait here. As long as I'm waiting, he'll definitely come. And even if I did go home, my heart would stay here behind this rock, waiting. My heart and my body would separate, and only my heart would freeze. I'm warmer if I stay here."

"Are you always waiting like this?"

"He told me to wait here every day, so I'm still waiting."

"No matter how many days you wait, it doesn't look like he's going to come. Your hands and feet must be frozen already. How about this—what if you were to plant a chrysanthemum in this rock, and let it wait here in your place?"

"I'll wait as long I'm alive. If I die here, a chrysanthemum will bloom here and wait in my place."

"He may not come even when there's a chrysanthemum waiting."

"He wants to come—there must be some reason why he can't. When I'm here where he told me to wait, somehow it seems as if he were already here. A chrysanthemum's color would never change, and it would go on blooming whether the person it waited for came or not. I'm the same."

"Your face has changed color already. You look as though you're going to freeze."

"If this autumn's chrysanthemum dies, another chrysanthemum will bloom next autumn. If a chrysanthemum takes my place, I'll be happy."

The ghost of the woman's head vanished, and now a chrysanthemum hovered in my dream. Snow began to fall on the rock. The rock grew white, the same white as the chrysanthemum, and it became difficult to distinguish the flower from it. Soon the snow and the rock and the chrysanthemum were lost in the ash-gray shroud of evening.

Thinking about the fact that a rock out in the middle of the mountains, in the middle of nature, had become the woman's grave, just as it was, I began to mutter to myself. Isn't that exactly what a seamless stupa is supposed to be? The woman's name wasn't carved anywhere on that large rock, or in the pool that lay in the shadow of the rock.

Long ago, in the Tang Dynasty, Dai Zong asked National Teacher Zhong of Nan Yang what he wanted when he died.

"Make this old priest a seamless stupa," Zhong is said to have answered. The seamless stupa has its origins in the section of *Blue Cliff Record* that describes this, "The National Teacher's Stupa."

A seamless stupa is a stupa with no seams, a stupa that cannot be perceived by the eye. Within this formless stupa all things in heaven and earth are contained. Thus the body of the stupa has the shape of an egg. The egg is a symbol of seamlessness.

When one sees the seamless stupas of succeeding generations of priests in the temple graveyard, it seems as though the priests are standing in a row with their round heads lined up.

But of course seamless stupas are made by people. They are stones made round, stones given the shape of an egg. Perhaps a truly seamless gravestone is a natural stone? The rock in my hometown and others of its sort are like this. Is that rock the grave of a woman with no grave? The woman didn't want the rock as a marker for her grave, and no one made the rock into a gravestone for her. A natural rock naturally became her gravestone. But then, do seamless gravestones really exist? There may be seamless lives, but I doubt that there are seamless gravestones. Perhaps the rock is a symbol of seamless life, and perhaps the white chrysanthemum that blossoms there is, too.

As long as there are flowers blooming in this world, as long as tall rocks stand against the sky, I do not need to have a stone carved for my grave. All of nature, all of heaven and earth, the old tale of the woman in the town of my birth, all of these things will be like gravestones of mine. I can only walk around looking at people's gravestones and considering them as works of art because I'm alive. What a foolish thing it is to dream of one's own gravestone. Thinking this, I walked out into downtown Kamakura, which was brilliant in the rays of the twilight sun.

First
Snow

on

Fuji

1.

"There's already snow on Mount Fuji. That's snow, isn't it?" said Jirō.

Utako looked out the train window at Fuji.

"You're right. The first snow."

"It isn't just clouds, is it? It's snow," Jirō repeated.

Mount Fuji was wrapped in clouds. The white clouds and the snow at the mountain's summit were nearly the same color, set against the overcast sky.

"What was today? September twenty-second?"

"Yes. Tomorrow is the middle of Higan. The equinox."

"I wonder if it doesn't snow on Mount Fuji every year around this time. Are you sure it's the first snow?" Jirō said. Then, as if the thought had just occurred to him, "Wait—you can't tell whether it's the first snow or not, can you? It's the first time you and I have seen snow on Mount Fuji this year, but it might have snowed before."

"Wasn't it in the paper? There was a big photograph—it said 'Mt. Fuji's First Facial of the Year.'"

"When was that?"

"I think it was this morning's paper. It wasn't last night's."

"I didn't see it."

"You didn't? You must not get the paper we get."

"Could be." Jirō smiled wryly.

"It looks exactly like the picture in the paper. I think it said it had been taken from the paper's airplane. Yes—the clouds were just like that. . . ."

Jirō remained silent, so Utako continued.

"The picture must have been taken yesterday if it was in this morning's paper. Yesterday's clouds were arranged just like these today. It's odd isn't it—the clouds keep moving, but the arrangement stays the same."

Jirō doubted that Utako had looked at the picture carefully enough that she could really say "yesterday's clouds were the same."

The fact that Utako had looked at Mount Fuji only after Jirō said "There's snow on Mount Fuji" was evidence of this. She hadn't noticed the mountain at all until he spoke. Surely Utako would have watched for the mountain when they boarded the train for Itō if it was true that the photograph had affected her so strongly. Surely she would have seen it before he did.

The train had already passed Ōiso.

No doubt Utako had remembered the picture in this morning's paper only after Jirō said "There's snow on Mount Fuji"—only after she looked at the mountain. Few people had reason to look so closely at a newspaper photograph of Mount Fuji.

If what Utako said was true—if the clouds around Mount Fuji had really been arranged in the same way yesterday as these today—it seemed to Jirō that one could sense in this the awesomeness of nature.

On the other hand, perhaps it was natural for Utako to forget the picture when she boarded the train with Jirō, even if it had affected her strongly this morning.

Utako had known that morning that she would take the train to Odawara with him and might well have made a mental note to mention the first snow once they reached a place from which Fuji could be seen. But even if this were the case, it was quite possible that in the end she had been so agitated she had forgotten to bring it up.

Utako and Jirō had been in love seven or eight years earlier, but she had married another man, and recently had gotten a divorce. Today she was going to Hakone with Jirō. She had many things to think about.

"It said in the paper that there would be snow as far as the eighth station on the path up the mountain—I wonder if that's the eighth station. . . ." Utako was still talking about Fuji's first snow. She glanced at Jirō's profile.

It seemed to Utako that for the first time since they'd met this morning Jirō's voice had bounded with its old liveliness again when he pointed out the snow on Mount Fuji. So she continued to speak of it.

His voice had sounded toneless and dull to Utako all the way from Tokyo Station, every time he answered her. She couldn't help wondering if he was depressed.

He was still looking out the window at Fuji.

Utako had grown very thin, and he was tempted to look at that haggard body of hers, to inspect it. This wasn't necessarily a cruel feeling—as a matter of fact it was love. Yet somehow the more he wanted to look at her, the less he was able to.

"As for what we were talking about before. . . ." Utako said.

She drew the conversation back from Fuji to her life.

"Mr. Someya, you mean?"

"Yes." For a moment Utako was silent. "I really think that for me, at this point—no matter what happens, as much as possible, I want to give people the benefit of the doubt."

"Yeah."

"If I keep resenting Someya, things will never get better for me."

"I think you're right."

"I mean, even our splitting up—maybe it was originally my fault. No, not just originally—if I really think about it, it seems like the whole thing was partially my fault."

"But if you're going to give people the benefit of the doubt, don't you think you should give yourself the benefit of the doubt, too?"

"Oh, of course. I suspect that the whole idea is a kind of trick to let me be nice to myself." Utako smiled.

As a young woman Utako's smile had been bright, but now it was lonely and twisted. One corner of her mouth bent up slightly, nervously.

"But that's not the only reason I want to think that way. It's also because I've gotten tired, that I don't have any energy. Maybe it's just easier to give people the benefit of the doubt when you're tired."

"Was your life together really such a battle?"

"Yes. When things start getting bad for a couple, there's really nothing you can do. So I guess I just did my best to put up with him. It's the woman who's in the house all the time, after all—it's the woman who's always having to endure . . ."

"Yeah, it certainly looks like it was a lot harder for you to break up with Someya. I guess it was nothing like when you and I broke up."

"Oh, that's cruel, saying that now. I didn't know anything then. Part of the reason I put up with so much this time was that I kept thinking about how you and I had split up before."

Jirō was unable to speak.

"Putting up with everything until we split up was much harder for me than actually splitting up. I really mean that."

Jirō nodded.

"And there are the children."

"You told me about the children earlier." Jirō had been looking at the snow on Mount Fuji, but now he looked back at Utako's face. "And since you've mentioned children—your children now, it looks like they'll grow up even if you're not there with them, doesn't it? But when you and I split up— because we split up—we killed that child." He spat these words out and then wished he hadn't said them.

Utako's lower eyelids and cheeks quivered. Even the tips of her fingers trembled.

"At that time—I didn't know anything about children then either."

Jirō saw that Utako's eyes had filled with tears.

"Yes, I guess that's true. More than anything else, the war was to blame. I really believe that," he said.

Utako shook her head.

"Getting pregnant made me get all confused. I got all confused—it got so I didn't understand anything."

Her eyes filled with tears again.

But Utako couldn't remember the child she had had with Jirō, the dead child. She could only remember the two children she had left at Someya's house.

"I agree that you were confused, that's true. Somehow—it was strange, but somehow your getting pregnant ended up splitting us apart," Jirō said.

Utako tried to forget her children with Someya for a few minutes, in order to remember her child with Jirō. But the baby had been taken from her soon after it was born, and she hadn't been able to ask where it had gone.

It was the year the war would end. Utako's parents guessed somehow that she was pregnant, then found out about her relationship with Jirō. Using the pregnancy as an excuse, the entire family fled Tokyo for a small town in the country. Since the region was new to them, they let it be known that their daughter was married, that they had brought her to the country to give birth—and that was that.

Utako's father was still working, so he'd remained for the most part in their house in Tokyo.

Her mother took her and the baby to Tokyo, plagued though the city was by air-raids. Utako held the baby in her arms. They went to get rid of the baby. Utako wanted to see Jirō, but she was taken back to the country the day after they gave the baby away.

Utako had not heard until after the end of the war that the child had died in the care of the person who took it.

"But—do you think the child really died?" Utako said.

Jirō looked away.

"Sometimes I think that it might still be alive, you know—possibly."

"I'm certain that it's dead."

"If it's alive, do you think if I met it somewhere—do you think I would know?"

"Look, the child is dead. Let's not talk about it."

Jirō had no desire to talk with Utako about anything in the past, not in detail. It wasn't just talk of the child.

2.

Utako's tears still weren't dry, so they took a taxi from Odawara Station. The corners of her eyes were red. It wasn't as though she had cried, she hadn't really, but she looked as though she had. Perhaps the constant strain on her mind and body had affected her eyelids. It seemed one had only to say something a little unpleasant and her eyes would fill with tears.

Jirō wanted to see Utako's face as it used to be. It was painful for him to look at her haggard features. And so from searching out the Utako he had known in the Utako before him, from trying not to see the Utako before him, his own eyes came to have a fatigued look to them. He didn't want her to feel that he was staring at her, but he didn't know where else to look.

Jirō hoped he might sense more of the old Utako in her features once they moved from the train to the taxi. She would probably look different in the cab, where it was just the two of them.

Jirō's desire to see the face he had known was so great that reasonings as minute as this came naturally to him.

When after time has passed a sound from long ago rings out a second time, joy and sorrow together become a poem. This was something a poet had written, but Jirō felt unconvinced. What exactly *was* a poem?

As the cab passed the castle remains at Odawara, Jirō looked at the grove of trees that grew there.

"Jirō—do you know who took the child?" Utako said quietly. In order to speak quietly she had drawn closer to him.

Jirō didn't know how to answer.

"Let's not talk about it."

"What? You knew?" Utako was surprised. "How did you find out?"

"Your father told me. I got a letter saying that the child had died."

"He——?"

"Probably he meant to say—'The ties that bound you have been cut'—I'm sure that's why he wrote. At the time I thought it might have been that he had lost his spirit—we were losing the war and all—and I thought maybe he felt guilty, that maybe that was why he let me know."

"You mean—my father told you?" Utako asked again, as though she was unable to believe what she had heard.

Then she leaned lightly against Jirō. Why she had drawn toward him like that he was unable to guess, whether it was because she had come to feel an intimacy of some sort between them, or because she no longer had the energy to support herself.

Feeling the warmth of Jirō's body, Utako let her eyelids drop.

Jirō waited for Utako to go on speaking, but she said nothing.

"It's all right if you lean on me," he mumbled.

Utako nodded, but she didn't come any closer. Indeed, her shoulders stiffened slightly, and she remained still.

"But even if my father told you, we don't know that what he told you was true. Being here with you like this, I feel sure that it can't be true," she said, speaking slowly, quietly.

It was like a lover's whisper. Utako's knees were almost trembling as she sat there pressed against Jirō. To control this trembling she tried to call up images of the children she had left at Someya's house, and spoke of the child she had conceived with Jirō.

Utako knew that Jirō pitied her. For this reason one part of her tried to hold back from him, to resist.

"I said this not too long ago, but—what he told me was definitely true," Jirō answered.

Jirō remembered receiving a letter saying that the child was dead and then going to see Utako's father. He had managed to get the address of the family that had taken the child, and he had even gone to the house to mourn. But he said nothing of this to Utako now.

"But I don't regret having had the baby then," Jirō said suddenly, his voice powerful.

Utako was surprised. It seemed she might pull away for a moment, but soon, as if agreeing, she moved closer.

"Even if it had a bad effect on your married life. . . ."

"There was nothing like that. That isn't true." Utako shook her head. "It wasn't like that."

They were leaving downtown Odawara, driving on a street lined with rows of cherry trees.

"At least it wasn't like that for Someya," Utako

said, correcting herself. "I don't think I would have come with you like this if it had been like that."

Jirō remained silent even when they passed the spa at Yūmoto.

It took unexpectedly little time to go from Miyanoshita to Kōran by car.

"Last time I came I took the train, and I remember it taking a really long time. But it was summer then, and there were these magnificent hydrangeas growing everywhere at the station—it was beautiful," said Jirō.

"Did you see the amaryllis blooming on the street back there?" Utako asked.

In Kōran there were a number of inns that had been set up after the war, converted villas of the so-called zaibatsu. The inn they had come to was one of these—one that had the remains of an old forest in its garden. Considering their style, the houses hardly looked as though they belonged to an inn.

The builders of the villa had hesitated to cut down trees that had been growing since the days when a natural forest covered the land. The room to which Utako and Jirō were led was shaded by leaves.

Neither of them knew the names of the trees, but sitting and looking at the trunks jutting up next to the verandah helped them to relax.

"It's nice here, isn't it? Just like a dream." Utako felt the tension rising from her as she looked at Jirō's face. "No—it's not like a dream, it's like I've woken up from a bad dream. It was a horrible way to live."

"We picked a nice place, didn't we?" Jirō spoke casually.

"Yes, there are nice places after all." As she gazed out at the numerous rocks in the garden, Utako thought how much she would like to bring her two children here. Even if she were going to part with them, it would be nice to bring them to a place like this and let them play at their leisure for a day. Then she could say good-bye.

"When my house burned in the air raids I went and rented a room in a temple in the country, out in Musashino. There was a man who taught Noh chanting who'd also had to leave his home, he'd put some tatami in the storeroom on the other side of the garden and was living there, and every once in a while drummers and someone who played the flute would come. Whenever I heard those drums and the flute I thought of you. I couldn't bear it."

Utako's eyes showed that she was pleased.

"Was your mother with you then?"

"Yeah, all three of us were there—my mother, my sister, and me."

"When did your sister marry?"

"I suppose it's been about four years."

Utako wondered when Jirō had gotten married— she still hadn't asked. She intended to say nothing at all about his wife.

Jirō continued his story. "The priest at the temple was also learning to chant in the Noh style—the teacher must have been counting on that when he came—and when I praised his chanting he said that he was awful at it, that he always ended up using the voice he used to chant sutras. . . . I'd hear him calling out *yō* and *hō* and so on, and the ponk of the higher drum being hit—I was always so startled my heart leapt. I was broken-hearted, and then on top of that I was malnourished, my body was weak. It struck me as peculiar and also as pretty amazing that they would go on hitting drums and playing flutes even as we were losing the war, you know? I mean—there probably wasn't anything else they could do, but still. . . . You and I didn't even have enough willpower left to think like that—to realize that there was nothing left for us to do but play our flutes. When the country lost the war you and I lost too."

"I was still a child, I didn't understand anything," Utako repeated. "But I think you're right—you and I should have been playing our flutes together. Things ended up like this because we weren't."

The maid came in again to suggest that they go along to the bath. It was the second time she had come.

"I've just checked the water, so feel free. . . ." she said.

"Yes, thank you. But I'm afraid we haven't brought wash cloths."

"I'll bring some to the bath."

Utako spoke after the maid left.

"How embarrassing—not bringing wash cloths. They'll wonder." Her face turned red.

They hadn't intended to come to Hakone when they met.

They had met in Ginza and eaten a late lunch, and afterward Jirō had walked with Utako to Shinbashi Station. He had looked up at the timetable for the Tōkaidō line while Utako was purchasing her ticket.

"Why don't we go to Hakone?" he had said suddenly.

"Today? Right now?"

She seemed to have been caught off guard.

Jirō hadn't had any dark ulterior motives in suggesting that they go to Hakone, certainly nothing to make her stiffen like that.

It was just that Utako was so terribly thin, and the way she fidgeted, as if she was frightened of something. One could see in her face how much her nerves had been worn down, and Jirō hadn't been able to bring himself to say good-bye to her there.

But if they entered the bath he would have to look directly at Utako's changed, worn body. At her body, scarred terribly by seven or eight years of marriage.

Utako still hadn't changed into the yukata the inn had provided when Jirō left for the baths. She hadn't even removed her socks.

3.

Jirō didn't really feel like washing himself in the sulphur spring, so having warmed himself in the water he sat on the edge of the bath and allowed his thoughts to drift. The hot water that came from the faucets in the area used for washing appeared to be fresh, but he didn't feel like using the soap the inn provided.

"Is it all right if I come in?" Utako said.

"Of course," Jirō answered. She slid the door to the room open just a little and stood holding it.

"When I was folding your shirt and things the maid came in and said—Oh, let me do that for you, you go join your husband in the bath. I really don't like it when she does that."

Utako was still wearing her light brown dress. She carried her yukata under her arm.

She looked quite nonchalantly at Jirō, who was naked. This wasn't at all what Jirō had expected.

"It's a bath. They won't let you get away without bathing."

"Really." Once she had closed the door, Utako hardly hesitated to come into the bath.

Jirō got no more than a glimpse of the color of her skin before he averted his eyes. The whiteness of her skin was beautiful.

Utako sat very still, sunk up to her neck in the water.

Jirō sat facing the same direction, looking out at the white flowers of a bush clover that spilled over the rocks near the window.

Utako lifted her shoulders from the water.

"I really thought it was strange, to tell the truth. We didn't run into one another once the whole time I was with Someya, and then the moment he and I split up we run into one another like this. I started wondering if this sort of thing really happens in the world. I started thinking that it must have been divinely arranged," she said brightly. "I mean, you were living in Tokyo too, right? And however huge Tokyo is, in seven or eight years you'd think we'd meet somewhere."

"But who knows? Maybe we passed one another on opposite sides of the street and just didn't notice, or something like that. Or maybe one person did notice but pretended not to, just kept walking, or maybe turned off onto a side street or something. . . ."

"What? What do you mean 'one person'? You mean you? Me?"

"I'm not saying it was like that with us."

"Well anyway—I almost never went out. Or rather, as a woman with small children I couldn't go out," Utako corrected herself.

She remembered worrying when she married Someya that she would run into Jirō somewhere, and wondering what she would do if she did.

Jirō remembered seeing women in the crowded trams and buses at the end of the war, packed in among those fleeing the air raids, whose profiles or whose figures seen from behind looked just like Utako's, and feeling as though a needle had been pushed through his heart. He couldn't say how many times this had happened, though he had known all along that Utako's family had been forced to evacuate and was living in the country.

"When you meet people, you always meet them in the most ordinary places. I kept thinking that if I met you it would be in some really wonderful location. But everyone who saw us was laughing, right?—because we crashed into one another. I doubt it looked very much like a meeting between people who hadn't seen each other for seven or eight years." Utako laughed.

They had met on the platform at Shinbashi station. Utako had already reached the top of the stairs when she saw that a man who resembled Jirō was getting on the train, so she hurried towards the door he was stepping through, they saw one another, and Jirō jumped down from the train just as Utako went to climb aboard—and there in front of the door of the train their bodies slammed together, just as the door slid shut.

The day they met they had arranged to meet again today.

"I've gotten skinny. Look at me." Utako laid a hand on the bone above her chest. "Even so, I look better than I did before I went back to my parents' house."

"Really?"

Jirō felt the old tenderness welling up within him now that they were in the water—the tenderness men feel toward women who have born their children—yet at the same time it almost seemed that the skin he saw was the skin of a new woman, and he felt confused.

"I lost a lot of weight after we split up, too—when the child died. But it wasn't like this. After all, I was young then."

Jirō realized that he didn't remember Utako's body very clearly, though he had always thought he did.

"I was young, and the way things were at the time it felt like I was the only person doing anything wrong, like I'd sinned. That's why I gave up the idea of living with you. I really think that's what it was. The war split up a lot of lovers, a lot of husbands and wives."

Utako had been conscripted to work in a factory making weapons. The feeling that everyone was criticizing her, the humiliation she had felt going into work pregnant day after day—it had all been so horrible she could hardly believe it when she remembered it now.

"Even marrying Someya—that was the fault of the war. I had no idea what was what then." Utako's eyes filled with tears again. "All I have to do is talk about these things and my heart starts pounding, that's how it is these days. Whenever Someya hit me or whenever we had an argument my chest would start to thump—it hurt so much I really thought I'd die if things kept going on as they were." Utako climbed out of the water, pressing a hand to her chest, and sat down at a faucet.

"Our youth was crushed by the war. But for

me—at least I had known you. I know I made you suffer, but—"

"No, that isn't true."

"You said you want to give people the benefit of the doubt, didn't you?"

"Yes. I realized how weak I had gotten when I went back to my parents' house, and it occurred to me that I had to look at things that way—that I had no hope of getting better if I didn't."

"I really hated you sometimes, and then sometimes I blamed myself for what had happened. And then I noticed that I was feeling a real nostalgia for my youth, probably because the lives of Japanese at that time were so miserable. I kept thinking that even in the middle of a war as awful as that one— even then I had had a lover like you. I guess you could say I was clinging to you."

"Well—it's nice to hear that."

They stood side by side, drying themselves.

Suddenly Jirō felt an urge to look at Utako's back without her knowing. It seemed strange to him that she wasn't curious about his body, that she gave no sign of wanting to look at him. Perhaps it was her feminine modesty, the docility of women that makes it possible for them to return to the past.

Utako's intimacy with him once she entered the bath affected Jirō, and dinner was enjoyable in a quiet way.

They had a six-mat and a three-mat room. They carried the low dinner table into the three-mat room, and as soon as the maid had finished preparing their bedding they lay down. It was still early.

"Let's talk until morning," Utako whispered. "But I don't want to talk about some things, okay?"

Jirō put his arm around her and drew her closer to him.

"Are you able to sleep well these days?"

"Oh, I'm always so exhausted. . . . "

Jirō wasn't sure whether she slept well because she was so exhausted, or was so exhausted she couldn't sleep.

"Hold me the way you used to," Utako said, lying very still.

"Hmm—how was that?" Jirō felt slightly at a loss.

Utako smiled. "I can't believe it—you've forgotten, haven't you?"

"And you used to be so quiet."

"Of course. I didn't know anything then."

Jirō closed his eyes, trying to call up images of Tokyo's streets burning in an air raid. He remembered the broken corpses. This was the method he used to keep his desires in check.

He used the same method whenever his wife felt ill, and he'd found it effective. Once, soon after the end of the war, Jirō had gone with a friend to a disreputable place, and the woman there had started talking about how she lost her family in an air raid. Jirō had paid no attention to her. The woman saw that Jirō didn't believe her, and began describing the corpses over and over. Jirō didn't doubt that she had seen the things she described, but even so, the people hadn't necessarily been related to her. Yet images of corpses he had seen began to rise in his own head.

"What's wrong?" the woman had said.

Jirō had replied curtly, "I'm allergic to war."

His method succeeded once again now, as he held Utako the way he used to.

Utako felt for Jirō's cheek in the dark, as if she too were wondering—what's wrong?

"What are you thinking about?"

"Unpleasant things from the war."

Utako wondered if he wasn't remembering his wife.

Jirō stroked Utako's hair gently.

Coming suddenly to Hakone, lying together at night like this—it all felt entirely natural to Jirō, as if it had all been planned out. Perhaps it was because Utako was being so docile. Although there was no doubt in his mind that her quietness was due to the fact that she was exhausted, terribly wounded.

"I probably would have been with you like this the whole time if there hadn't been a war. Don't you think so?"

"But we met at the factory. You would never have gone the factory if there hadn't been a war."

"Even if we hadn't met at the factory, though—I think we would have met somewhere. I'm sure we would have."

Jirō was conscious that the scent of Utako's hair was unique, that it was a scent no other woman had.

What had changed in the quiet young woman he had known? How had she changed during her seven or eight years of marriage, in giving birth to two children? Jirō felt jealousy and temptation, but he filled his head again with images of the war's broken corpses.

Utako's body was so skinny that he hadn't been able to leave her at the station, and now it had come to this. Jirō muttered silently to himself that her thinness was his own responsibility, too. He tried to convince himself that the reason he held her was not that he felt a new burgeoning of desire.

Even assuming that he felt no desire, the near miraculous efficacy of imagining the war's corpses struck him as frightening.

Utako was leaving everything to Jirō, she was flexible—but then too, he could feel the strength draining from her body in his hands.

Utako felt relaxed, it was true. Yet at the same time she felt a loneliness, like a dying flame.

She had caught her breath when, at Shinbashi station, Jirō suddenly suggested they go to Hakone—had that catching of breath been pointless? At the time she had resolved to resist him as much as she could—the thought had come to her suddenly, it had flickered into her head—but thinking back now it seemed pathetic even to have thought such a thing.

Utako lay still for a short time, then began to sob. She moved her face closer to Jirō's. He was surprised to see that her cheeks were soaked with tears. He wiped her cheeks with the palm of his hand.

"I do cry a lot, don't I." Utako laughed. "My parents are amazed."

"Yeah, your nerves are really worn down. Divorce is a tough thing for anyone to have to go through."

"That's not true. Didn't I tell you before that it's everything you have to go through until you split up that's hard? All the putting up with things? It's so hard that when all the ropes finally come untied, it's like your body is floating through space."

"It was my fault that your marriage went so badly. I was praying for your happiness, you know—off in the shadows. But I was too complacent. I should have been much more critical of myself."

"It really wasn't your fault, Jirō. I said there were some things I didn't want to talk about, but—do you mind if I talk a little about the way things were before Someya and I split up?" Utako felt about for Jirō's hand. "I never even dreamed that I would have an opportunity to talk to you about all this. I never thought I'd see you again."

4.

Utako was sleeping with her back to him when Jirō woke up the next morning. Her legs were bent up slightly toward her chest.

Gazing from behind at her rather childlike form as she slept, Jirō felt a tiny smile spring to his lips. He put out a hand and lightly touched her hair.

Utako turned to face him. Jirō was surprised at her sensitivity and drew his hand back, but she hadn't awakened.

There were no gaps between the sliding storm doors, so the room was only dimly lit. Jirō lay looking at Utako's face, and a love like that he had felt in the old days came welling up within him. He had the impression that her face hadn't changed at all.

Jirō shut his eyes, but he didn't feel like going back to sleep, so he got up and went to the bath.

Utako was lying awake in bed when he returned.

"Have you been to the bath already? You didn't wake me up?"

"It's nine, you know."

"Nine? . . . How embarrassing. I haven't slept so well in ages."

"Well, that's good. And last night, too—you were asleep before I was, weren't you? It must have been about twelve."

"Nine hours. Oh, I feel so good. . . ."

Utako was savoring her comfort, and took her time getting out of bed.

"You were sleeping with your back to me, all scrunched up."

"Was I? . . ."

"Perhaps it's a habit. You must sleep with your back to Someya."

"I wonder." Utako sat up and stared at Jirō.

It was a long time before she returned from the bath.

Jirō walked in the garden while the maid was cleaning the room.

Leaning against the trunk of a tall tree, he addressed Utako. She sat with the mirror in front of her, doing her makeup.

"How would you like to go to Ashi-no-ko?"

"Ashi-no-ko? . . ."

"Fuji and its first snow might be reflected in the lake. The weather is certainly nice enough."

"And it's the equinox."

"From what I've heard, you take a cable car up from here and go by bus to the bottom of the lake. You take a tour boat from there."

"Really?" Utako poked her face out from behind the mirror. "Are you going? I don't want to move. I just want to stay here and rest."

"Well then, let's not go."

Jirō stepped up into the room.

"You certainly had a long bath, didn't you?"

"There aren't very many spas that you can see mountains from—I just sat there gazing out at the mountains. I was wondering what it would have been like if I had come here with you in the old days. I tried pretending that I had come with you then."

"I see," Jirō said, nodding. "Though in those days a man couldn't really go to a spa with a woman, could he?"

"And now it's just being cared for—being comforted."

Jirō was unable to respond.

"But that's okay. Different things are necessary for people at different times—I was thinking that earlier. And what's necessary for me right now is care and comforting."

They ate breakfast quietly, taking their time.

Utako served the food after the maid withdrew. The natural intimacy of this seemed almost strange to Jirō.

Jirō had been disturbed by those words of Utako's, certainly, but it wasn't as though they had passed the night as they had because he felt disillusioned by the skinniness of her body, and he wasn't at all afraid that there would be troublesome complications. At least—he couldn't say for certain that any of this was true, but he didn't think it was.

Of course if he had come with a woman he had only met recently and spent a night like that, this morning would undoubtedly have been awkward. Certainly there wouldn't have been the intimacy he felt with Utako.

But this too was difficult to say.

"When we split up back then, you know—I really thought it was over. I gave up completely. But it looks like there was still something important there between us. I hope we can take care of what's important."

"You're talking in riddles."

"It's like a riddle."

"A riddle that can't be solved? Or a riddle that can be solved?" Utako tilted her head slightly to one side, as if she were trying to decide which it was.

"When two people meet again after so many years, after having broken up—just learning that they didn't hate one another is enough to make them happy. Unbelievably happy."

"It's true, isn't it?"

They boarded a bus that left a little after two and got off at Odawara.

Riding on a train bound for Tokyo—heading in the opposite direction from the day before—they looked once more at the first snow on Fuji.

"There are no clouds at all today. You can see all the way to the foot of the mountain."

"It's not very interesting, is it—there's really just the tiniest bit of snow at the summit. You can see that now that the clouds are gone."

"Oh, I don't know. . . ." said Utako, casually touching Jirō's hand. "You don't think it's because we saw it yesterday? Even looking at Mount Fuji must get boring if you see it all the time."

Jirō understood that Utako was ready to say good-bye.

"Thank you very much for bringing me. I really had a nice time. And I have a feeling I may get better now."

Utako had put all her strength into those words. Jirō's face clouded.

"I really mean that," Utako said, taking Jirō's hands between the palms of her own.

Jirō went on looking at the first snow on Fuji.

Silence

It is said that Omiya Akifusa will never say a word again. It is said that he will never again write a character—though he is a novelist, and only sixty-six years old. What is meant by this is not simply that he will no longer write novels, but that he will no longer write even a single letter or character.

Akifusa's right hand is paralyzed, is as useless as his tongue. But I have heard that he can move his left hand a little, so I find it reasonable to assume that he could write if he wanted to. Even granting that he would find it impossible to write passages of any length, still it seems likely that he could write words in large *katakana* when he wanted to ask that something be done for him. And since he is now unable to speak—since he can neither signal nor gesture freely—writing even the most crooked *katakana* would allow him to communicate his thoughts and emotions in a way not otherwise possible. Certainly misunderstandings would be less common.

However ambiguous words may be, they are certainly much easier to understand than clumsy body language or awkward gestures. Even supposing that old Akifusa managed to show that he wanted something to drink—by pinching his lips into a shape that suggested sucking, for example, or by miming the act of lifting a cup to his mouth—just making it clear whether it was water or tea or milk or medicine that he wanted, which of just these four—even that would prove difficult. How would he distinguish between water and tea? It would be perfectly clear which he meant if he wrote "water" or "tea." Even the single letter "w" or "t" would get the message across.

It is strange, isn't it, that a man who has made his living for forty years using letters and characters to write words should, now that he has almost entirely lost those letters and characters, and consequently come to understand the powers they possess in the

most fundamental sense, and with the greatest cer-
tainty—now that he has become able to use them
with such knowledge—it is strange, is it not, that he
should deny himself their use. The single letter "w"
or "t" might be worth more than all the flood, the
truly tremendous flood of words and letters he has
written in his life. That single letter might be a more
eloquent statement, a more important work. It
might well have more force.

I thought I might try saying this to old Akifusa
when I visited him.

Going from Kamakura to Zushi by car, one passes
through a tunnel, and the road is not very pleasant.
There's a crematorium just before the tunnel, and
it's rumored that lately a ghost has been appearing
there. The ghost of a young woman shows up riding
in cars that pass beneath the crematorium at night—
so the story goes.

It would still be light when we passed, so there
was nothing to worry about, but nonetheless I asked
the familiar-looking taxi driver what he knew.

"I haven't encountered her yet myself, no—but
there is one fellow in our company who's given her
a ride. And it isn't just our company, either—she's
ridden in other companies' cabs too. We've got it
arranged so we take a helper along if we take this
road at night," the driver said. Judging from the way
he spoke, he had repeated the story often enough to
make him tired of telling it.

"Where does she appear?"

"Where indeed. It's always cars coming back empty from Zushi."

"She doesn't appear when there are people in the car?"

"Well—what I've heard is that it's empty cars coming back. She fades in near the crematorium, I guess. And from what I hear it's not like you stop the car and she gets in, either. You don't know when she gets in. The driver starts feeling a little weird and turns around, and there's this young woman in the cab. But since she's a ghost there's no reflection of her in the rear-view mirror."

"That's bizarre. I guess ghosts don't reflect in mirrors?"

"I guess not. They say she has no reflection. Even if human eyes can see her . . . "

"Yes, but I suppose human eyes would see her, wouldn't they? Mirrors aren't quite so impressionable," I said. But of course the eyes looking at the mirror were human eyes, weren't they?

"But it isn't just one or two people who've seen her," said the driver.

"How far does she ride?"

"Well, you get scared and kind of dazed, and so you start driving really fast, and then when you come into downtown Kamakura you relax, and by then she's already gone."

"She must be from Kamakura, then. She must

want to go back to her house in Kamakura. They don't know who she is?"

"Ah, now that I don't know. . . ."

Even if he did know, or if there was some talk among taxi drivers about who she might be and where she might come from, it was doubtful that the driver would be careless enough to say so to a passenger.

"She wears a kimono, the ghost—she's quite a beautiful woman. Not that anyone's looking back over their shoulder at her or anything. You don't exactly ogle a ghost's face."

"Does she ever say anything?"

"I've heard she doesn't speak. It'd be nice if she'd say thank you at least, right? But of course when ghosts talk they're always complaining."

Just before we entered the tunnel, I looked up over my shoulder at the mountain on which the crematorium stood. It was the Kamakura City Crematorium, so it seemed that most of the dead cremated there would want to return to houses in Kamakura. Maybe it would be nice to have a woman as a symbol of all those dead, riding around at night in empty cabs. But I didn't believe the story.

"I wouldn't have thought that a ghost would need to take a cab. Can't they go wherever they like, appear anywhere?"

We arrived at Omiya Akifusa's house soon after we exited the tunnel.

The cloudy four o'clock sky was faintly tinged with peach—a sign that spring was coming. I hesitated for a moment at the gate of the Omiya house.

I had only been to visit old Akifusa twice in the eight months since he had himself become a sort of living ghost. The first time was just after his stroke. He was some twenty years my senior, a man I respected, a writer who had been a patron of mine—it was hard for me to see him like that, ugly and miserable as he had become.

But I knew that if Akifusa ever had a second stroke, it was likely to be the end. We lived in neighboring towns—Zushi and Kamakura are no more than a stone's throw apart, as we say—and the fact that I had neglected to visit had begun to trouble my conscience. The number of people who had left this world while I considered visiting them, but somehow never did, was by no means small. It had happened so often I had come to believe that this was simply the way life was. I had been thinking of asking Akifusa to write out something for me on a sheet of *hansetsu* paper, but the idea had come to seem pointless. And the same thing had happened to me several times. Sudden death wasn't the sort of thing I could treat casually. I was perfectly aware that I myself might die at any moment—perhaps one night in a storm—and I did not take good care of myself.

I knew other authors who had died suddenly of cerebral hemorrhages or heart failures, of coronary strictures—but I had never heard of anyone being saved but paralyzed, as old Akifusa had been. If one views death as the greatest misfortune of all, one would have to say that Akifusa was extremely fortunate to go on living, even though he lived as a patient with no hope of recovery—as a disabled man. But that sense of good fortune was rather difficult for most of us to feel. It was also difficult for us to tell whether Akifusa himself felt that he had been fortunate.

Only eight months had passed since Akifusa suffered his stroke, but from what I'd heard the number of people visiting him had already become quite small. It can be difficult to deal with an old deaf man, but it's no less difficult to deal with a man who is able to hear but can't speak. He understands what you say to him, though you don't understand what he wants to say to you—it's even stranger than talking to a deaf person.

Akifusa's wife had died early on, but his daughter, Tomiko, had remained with him. There were two daughters, but the younger one had married and moved out—Tomiko, the older daughter, had ended up staying to take care of her father. There was no real necessity for Akifusa to remarry, since Tomiko took care of all the household chores—indeed, he had relished the freedoms of single life—and one might say for this reason that Tomiko had been obliged to make certain sacrifices for her father. The fact that Akifusa remained single, despite having had numerous affairs, suggests too that the power of his will was such that it overcame his emotions—or perhaps something else was going on.

The younger daughter was tall and had exceptionally fine features—she resembled her father somewhat more than her sister did—but this wasn't to say that Tomiko was the kind of young woman one would expect to remain unwed. Of course she was no longer a young woman—she was approaching forty, so she wore almost no makeup at all—but one sensed a purity in her. She seemed always to have been a quiet sort of person, but she had none of the gloominess and irritability of the proverbial old maid. Perhaps her devotion to her father provided some comfort to her.

People who came to visit always talked to Tomiko instead of Akifusa. She sat beside her father's pillow.

I was startled to see how terribly thin she had become. It seemed odd to me that I was surprised, which meant that it was natural for her to be thin—but even so, seeing Tomiko suddenly grown so old and shriveled depressed me. It occurred to me that the people in that house were suffering.

There was nothing for me to say once I had said the pointless words of a sickbed visitor, so I ended up saying something I shouldn't have said.

"There's a rumor going around that a ghost has been appearing on the other side of the tunnel—have you heard? I asked the driver about it on the way here, as a matter of fact. . . ."

"Is there really? I'm always in the house—I don't hear about anything." It was clear that she wanted to know more, and so—thinking all the while that it would have been better not to mention the matter—I summed up what I knew.

"Well, it's the sort of story one can hardly believe—at least not until one actually sees the ghost. Of course, one might not believe the story even if one did see it. There are always illusions, after all."

"You ought to look for it tonight, Mr. Mita—find out if it really exists," Tomiko said. It was an odd thing for her to say.

"Oh, but you see it doesn't appear when it's light out."

"The sun will have set if you stay for dinner."

"Unfortunately I'm afraid I can't. Besides, it seems that the woman's ghost only rides in empty cars."

"Well if that's the case you have nothing to worry about, do you? My father is saying that he's extremely pleased you've come, and that he hopes you'll make yourself at home. Father—you'll have Mr. Mita to dinner, won't you?"

I looked at old Akifusa. It seemed as though the old man had nodded his head on his pillow. Was he pleased that I had come? The whites of his eyes were clouded and bleary, and there were smudges of yellow even in his pupils, but it seemed that from the depths of those smudges his pupils were glittering. It seemed that he would suffer his second stroke when those glitterings burst into flame—it seemed it might happen at any moment—and I felt uneasy.

"I'm afraid I'll tire your father if I stay too long, and that might . . . "

"No. My father doesn't get tired," Tomiko stated clearly. "I realize that it's unpleasant of me to keep you here with a man as sick as my father, but he remembers that he's a writer himself when there's another writer here. . . ."

"He—what?"

I was somewhat surprised by the change that had occurred in Tomiko's tone of voice, but I prepared myself to stay for a while.

"Surely your father is always aware that he's a writer."

"There's a novel of my father's that I've been thinking about a lot since he became like this. He wrote about this young man who wanted to be a writer—the boy had been sending strange letters to him pretty much every day, and then he went completely mad and was sent off to a sanatorium. Pens and inkpots are dangerous, and they said that pencils were dangerous too, so they wouldn't let him have them. Manuscript paper was the only thing they would let him have in his room. Apparently he was always there in front of that paper, writing . . . at least he thought he was writing. But the paper stayed white. That much is true, the rest is my father's novel. Every time the boy's mother came to visit, he would say—Mom, I wrote it, Mom, will you read it? Mom, will you read it to me? His mother would look at the manuscript he handed her and there would be nothing written on it at all, and she would feel like crying, but she'd say—Oh, you've written it very well, it's very good, isn't it!— and she would smile. Every single time she went he would pester her to read the manuscript to him, so she starts reading the blank paper to him. It occurs to her to tell him stories of her own, making it seem that she's reading the manuscript. That's the main idea behind my father's novel. The mother tells the boy about his childhood. No doubt the crazy boy thinks he's having his mother read some sort of record of his memories, something that he wrote himself—that's what he thinks he's listening to. His eyes sparkle with pride. His mother has no idea whether or not he understands what she's saying, but every time she comes to see him she repeats the same story, and she gets better and better at telling it—it begins to seem like she's actually reading a story of her son's. She remembers things she had forgotten. And the son's memories grow more beautiful. The son is drawing the mother's story out, helping her, changing the story—there's no way of telling whose novel it is, whether it's the mother's or the son's. When the mother is talking she's so focused she forgets herself. She's able to forget that her son is mad. As long as her son is listening to her with that complete concentration, there's no way of knowing if he's mad or not—he could very well be mad and sane both. And at those times the souls of the mother and the child fuse together—it's like the two of them are living in heaven—and the mother and the child are both happy. As she goes on reading to him it begins to seem that her son might get better, and so the mother goes on reading the blank paper."

"That's the one called *What a Mother Can Read*, isn't it?—one of your father's masterpieces. An unforgettable work."

"The book is written in the first person—the son is the 'I'—but some of the things he remembers about his childhood actually happened to my sister and me when we were small. He just had it all happen to a boy. . . ."

"Is that so?"

It was the first time I had heard this.

"I really wonder why my father wrote a novel like that. The book scares me now—now that he's like this. My father isn't mad, and I can't be like that mother and read him a novel that hasn't been written down—but I do wonder if he isn't writing a novel in his head even now."

It struck me that Tomiko was a peculiar person—able to say things like this even though old Akifusa probably heard every word. I didn't know how to respond.

"But your father has already written numerous outstanding works—he and that literary-minded boy are entirely different."

"Do you think so? I think that my father still wants to write."

"Of course, not everyone would agree."

Personally I thought he had written quite enough already, but if I were in old Akifusa's condition, I had no idea what I'd think.

"It's just that I can't write for my father. It would be nice if I could write *What a Daughter Can Read*, but I can't. . . ."

Her voice sounded to me like the voice of a young woman in hell. The fact that Tomiko had turned into the sort of woman who said such things—could it mean that by being in constant attendance on her father, who was a sort of living ghost, she had been possessed by something in him? It occurred to me that she might write a book of horrifying memories when Akifusa died. I began to feel a powerful hatred.

"What if you were to try writing about your father. . . . "

I refrained from adding—while he's still alive. Suddenly I remembered some words of Marcel Proust's. A certain nobleman has abused lots of people in his memoirs, which are at long last about to be published, so he writes, "I'm on the verge of death. I hope my name doesn't get dragged around in the mud too much, since I'll be unable to answer." Not that it was at all like that with Akifusa and Tomiko. They were by no means strangers—indeed, there may have occurred between them a mysterious or perhaps a perverted emotional interchange, something beyond what most fathers and daughters experience.

I was struck by the strange thought that Tomiko might write about her father as if she had become her father.

Whether it became an empty game or a moving work of art, it seemed that either way it might provide some comfort for both of them. Akifusa might be saved from his absolute silence, from verbal starvation. Verbal starvation is surely something intolerable.

"Your father would be able to understand what you wrote, and he'd be able to evaluate it—you wouldn't be reading a blank sheet of paper, and if you really wrote about your father, if you had him listen to you read . . . "

"Do you think it would be my father's work? If even a little of it . . . "

"There's no question that some of it would, at least. Anything more—it's up to the gods, and it also depends on how close the two of you are. I have no way of knowing."

But it did seem that a book written in such a way would have more life than a book of memories written after Akifusa was dead. If it went well, even the sort of life he was living now could be preciously literary.

"Even if your father goes on being silent, he could still help you, and he could still fix your mistakes."

"It wouldn't have any meaning if it ended up being my own work. I'll have to talk it over very carefully with my father." Tomiko's voice was lively.

Once more I seemed to have said too much. Wasn't what I was doing like forcing a desperately wounded soldier to return to battle? Wasn't it like violating a sanctuary of silence? It wasn't as though Akifusa was unable to write—he could write letters or characters if he wanted to. Perhaps he had chosen to remain silent, chosen to be wordless because of some deep sorrow, some regret. Hadn't my own experience taught me that no word can say as much as silence?

But if Akifusa was to continue in silence—if his words were to come from Tomiko—wouldn't that be one of the powers of silence, too? If one uses no words oneself, other people speak in one's place. Everything speaks.

"Shall I? My father says that I should give you some saké right away—that at the very least." Tomiko stood.

I looked instinctively at Akifusa, but there was nothing to suggest that the old man had spoken.

The two of us were alone now that Tomiko had left, so Akifusa turned his face in my direction. He looked gloomy—maybe there was something he wanted to say? Or maybe it irked him to be put in a situation where he felt as though he had to say something? I had no choice but to speak myself.

"What are your thoughts regarding what Tomiko was saying just now?"

". . . . ."

I addressed silence.

"I feel sure that you could produce an intriguing work, really quite different from your *What a Mother Can Read*. I started to feel that way as I was talking with Tomiko."

". . . . ."

"You never wrote an 'I Novel' or an autobiography—perhaps now that you yourself are no longer able to write, using some other person's hand to produce a work of that sort might allow you to reveal one of the destinies of art. I don't write about myself, and I don't think I could write about myself even if I tried, but if I were silent and if I could write like that . . . I don't know whether I'd feel a sort of joy, as though I had finally realized the truth—if I'd think, is this who I am?—or if I'd find the whole thing pathetic and give up. But either way, I'm sure it would be interesting."

". . . . ."

Tomiko returned with saké and snacks.

"Can I offer you a drink?"

"Thank you. I hope you'll forgive me for drinking in front of you, Mr. Omiya, but—well, thank you."

"Sick people like him don't make very good conversation, I'm afraid."

"I was continuing our discussion from before, actually."

"Were you? As a matter of fact, I was thinking as I was heating up the saké that it might be interesting if I was to write in my father's place about all the affairs he had in the years after my mother died. He told me everything about them in great detail, and there are even some things that my father has forgotten and I still remember. . . . I'm sure you're aware, Mr. Mita, that there were two women who rushed over here when my father collapsed."

"Yes."

"I don't know whether it's because my father has been in this condition for so long, or whether it's because I'm here, but the two women have stopped coming. I know all about them, though—my father told me all about them."

"But your father doesn't see things in the same way you do." This was obvious, but even so Tomiko seemed irritated.

"It's impossible for me to believe that my father has told me any lies, and it seems that over time I've come to understand his feelings. . . ." She stood up. "But why don't you ask him yourself. I'll get things ready for our dinner and then I'll be back."

"Please, don't worry about me."

I went along with Tomiko and borrowed a cup. It's best to get the saké in quickly when you're talking to a mute.

"It seems as though your love affairs have become Tomiko's property now. I guess that's the way the past works."

". . . . ."

I may have hesitated to use the word "death"—perhaps that was why I had said "the past."

But surely as long as he was alive the past was old Akifusa's property? Or should one think of it as a sort of joint ownership?

"Maybe if it were possible for us to give our past to someone, we'd just want to go ahead and give it."

". . . . ."

"A past really isn't the sort of thing that belongs to anyone—maybe I'd say that one only owns the words that are used in the present to speak about the past. Not just one's own words—it doesn't matter whose words they are. No, hold on—except that the present instant is usually silent, isn't it? Even when people are talking like I am now, the present instant is just a sound—'I' or 'a' or 'm'—it's still just meaningless silence, isn't it?"

". . . . ."

"No. Silence is certainly not meaningless, as you yourself have. . . . I think that sometime before I die I would like to get inside silence, at least for a while."

". . . . ."

"I was thinking about this before I came, but—it seems that you should be able to write out *katakana* at least, and yet you refuse to write even a single letter. Don't you find this at all inconvenient? If there's something you want done—for example, if you wrote 'w' for water or 't' for tea . . . "

". . . . ."

"Is there some profound reason for your refusal to write?"

". . . . ."

"Oh—I see now. If the single letters 'w' and 't' and so on are enough to get things done, the sounds 'I' and 'a' and 'm' must not be meaningless either. It's the same with baby talk. The baby understands that its mother loves it. That's how it is in your *What a Mother Can Read*, isn't it? Words have their origin in baby talk, so words have their origin in love. If you were to decide to write 't' every time you wanted to say thank you—and if every once in a while you wrote 't' for Tomiko—just think how happy she would be."

". . . . ."

"That single letter 't' would probably have more love in it than all the novels you've written during the past forty years, and it would probably have more power."

". . . . ."

"Why don't you speak? You could at least say 'aaa'—even if you drool. Why don't you practice writing 'a'?"

". . . . ."

I was at the point of calling into the kitchen to ask Tomiko to bring a pencil and some paper when I suddenly realized what I was doing.

"What am I doing? I'm afraid I've gotten a little drunk—forgive me."

". . . . ."

"Here you'd gone to all the effort of achieving silence, and then I come along and disturb you."

". . . . ."

Even after Tomiko returned I felt as though I had been babbling. All I had done was circle the perimeter of old Akifusa's silence.

Tomiko used the telephone at a nearby fish store to call the driver who had brought me.

"My father is saying that he hopes you'll come talk with him again from time to time."

"Yes, of course."

Having given her this rather offhand answer, I got into the car.

"Two of you have come, I see."

"It's still early in the evening and we do have a passenger, so I doubt she'll show up—but just in case. . . ."

We came out of the tunnel on the Kamakura side and drove under the crematorium. Suddenly, with a roar, the car began to fly.

"Is she here?"

"She's here. She's sitting next to you."

"What?"

The effects of the alcohol disappeared in a flash. I glanced to the side.

"Don't frighten me like that. I'm in no condition to deal with it."

"She's there. Right there."

"Liar. Slow down, will you—it's dangerous."

"She's sitting right next to you. Can't you see her?"

"No I can't. I am utterly unable to see her," I said. But as I said this I began to feel a chill. I tried to sound brave. "If she's really there—what do you think, shall I say something to her?"

"D—Don't even joke like that. You get cursed if you speak to a ghost. You'll be possessed. It's a terrifying idea—don't. Everything will be fine if we just keep quiet until we've taken her as far as Kamakura."

Her Husband Didn't

First her ear, then her eyebrow, then . . .
One by one the various parts of Kiriko's
already-married body drifted up before
Junji, filling his head. He was thinking his
way through the sequence of kisses he
would give her that evening.

It takes about an hour to go by train from
North Kamakura to Shinbashi on the Yokosuka line. There was enough time for him to

imagine a number of different sequences if he liked, and various methods of kissing.

Though Junji and Kiriko both lived in North Kamakura, they met in Tokyo as a rule—Kiriko worried that they would attract attention if they were seen together in their own small town. They took the added precaution of riding in on different trains. Each time they arranged to meet, Junji suggested that he be the one to take the earlier train and wait. Kiriko never even had to ask him to do this. Junji was young, still a student. He seemed to fear that he might discover some flaw in her body.

First the ear. . . . Junji started with Kiriko's ear because he still regretted the disappointment he had felt the first time he touched her earlobe— because he was still sorry that it had not excited him. His disappointment had been so acute on that occasion that even the color of his face must have changed.

"Hey." Kiriko had opened her eyes. "What's wrong?"

He had pulled his finger back from her ear the instant he had touched it. No doubt that, too, seemed strange.

Junji took Kiriko's ear hurriedly into his mouth. His face was hidden in the hair at the side of her head. The smell of hair engulfed him.

"I wish you wouldn't do that."

Junji caught hold of her head as she tried to pull away.

Kiriko's ear was small and soft but not fleshy, and Junji could fit all of it into his mouth. His disappointment vanished.

But the fact that Junji had felt an urge to touch Kiriko's earlobes at all—there was something in this that filled him with a sense of guilt. Because for Junji, this desire was linked to the abnormal excitement he had felt on a previous occasion, when he had fingered the earlobe of a prostitute.

Junji had scarcely known what he was doing when he had pinched that woman's earlobe—his first earlobe. It wasn't that he had admired its shape, and pinched it for that reason, but then what was it? Why, just when he was feeling such self-hatred, such a powerful aversion to the idea of touching any part of a woman at all—why had his hand moved to her ear? Junji himself couldn't say.

Yet the cold feeling of that earlobe had instantly cleansed him of his filth. The earlobe was just as round and plump as an earlobe ought to be—it was small enough that Junji could squeeze it between the tips of his thumb and forefinger, no bigger than that—yet it filled him with a sense of the beauty of life. The smooth skin, the gentle swelling—the woman's earlobe was like a mysterious jewel. Her purity had remained intact there, inside it. The earlobe held dew-like droplets of the essence of female beauty. A sentimentality like yearning welled up inside Junji. He had never known anything with a texture like this. It was like touching the lovely girl's soul.

"What on earth are you doing?" The woman shook her head cruelly.

Even after he'd left the woman's house, Junji said nothing to his friends about the ear. They would only have laughed at him if he had. And though it would be difficult to bring that sense of excitement to life within him again, ever again, it became Junji's secret—a secret that would probably stay with him for the rest of his life.

Still, when Junji considered that his desire to touch Kiriko's earlobe had originated in his memories of a prostitute's earlobe, he quite naturally suffered pangs of guilt.

Yet for all that, Kiriko's earlobe had betrayed his expectations. It felt thin and flimsy in his fingers. It lacked even the moisture and the silkiness of most earlobes—it seemed crusty and dry. Junji was startled. He was so confused it never even occurred to him that he might never feel the excitement he had felt touching that prostitute's earlobe again—that touching even the loveliest earlobes might not revive that emotion.

Junji's habit of kissing the various parts of Kiriko's body started to form the moment he took her ear in his mouth.

Until then things had been simple. He was still a beginner in the affairs of love, and he had been completely overwhelmed by the surprise he felt on discovering that he could satisfy the middle-aged Kiriko so completely. He came to understand his own masculine charm for the first time through the pleasure Kiriko took in him, and he drifted drunkenly in it.

Junji had believed that Kiriko's body was entirely perfect, so he had to distract himself instantly from what he felt in his fingers—the impoverished flimsiness of her earlobes. He was also conscious of the fact that the ecstatic joy Kiriko had experienced when she was with him at first was growing less ecstatic of late. It was as though the attempt to rekindle her passion had given him an excuse to touch her ear.

Then, during only the third or fourth of their rendezvous, Kiriko said something unexpected.

"Sometimes I used to wonder if I could go to bed with a man without having to worry about any of those difficult ties like marriage or love, you know? I wanted to try sometime—just once. I used to daydream about it."

It sounded to Junji as though that daydream had been realized through him—no other interpretation seemed possible. He felt as though he had been pushed off a cliff.

"You mean—this has all just been a game for you?"

Kiriko firmly denied this. "It's not a game. Men may play games like that, they may fool around and everything—but women aren't like that. At least I'm not like that."

"I don't see how you can say that. What you said just now—you could only have meant you think of this as a game. If it isn't a game . . . "

"I don't know how to explain. There's just something about it—there's this aura of secrecy," she mumbled. "You really don't have the slightest idea how many constraints and burdens a woman my age has to deal with, do you? And in that oppressive pain there's a secret. But I see now that it would have been better if it had ended as it was—as a secret daydream."

"You're sorry that you slept with me?"

Kiriko laughed at this trite, childish line.

"Asking me something like that—if I'm sorry I slept with you! Aren't you just insulting yourself? Even if I criticize myself and even if I feel pain—I still wouldn't want to say that I regret what I did. Regret is the easiest excuse of all. It's just a convenient way to escape. . . ."

"So that's all this was? I just happened to end up playing opposite you in these secret daydreams of yours?"

"You didn't find it at all strange that I started seeing you like that? It didn't seem strange that it all happened so easily? I had never had an affair, you know."

". . . . ."

"Think about it—I told you about my late daughter the very first time I met you. . . ."

This had happened on a train on the Yokosuka line. Junji had attended a class on Western-style painting that day on an invitation from a friend. They had learned to sketch female nudes. There were four or five young women in the class, but Kiriko was the only one in Japanese dress, and she was older than the other women, so she had attracted Junji's attention. It turned out that they both lived in North Kamakura, and so that evening they went back together on the train. When the conductor came to collect their tickets, Kiriko handed him the money necessary to change Junji's third-class ticket to second-class before Junji could get his money out. It was quicker for her to open the handbag on her lap than it was for Junji to rummage through his pocket, but her movements suggested that she had been planning to pay.

Sometime after they passed Yokohama, Kiriko opened her sketchbook and started to draw something. It seemed to Junji that her face became more and more beautiful as she alternately glanced up at him and looked down at the paper. They were sitting across from one another. Junji leaned forward and looked at the sketchbook—Kiriko was sketching his face. He took the sketchbook from her without saying a word. He looked at it for a moment or two, then took out his own pencil and began adding to the sketch she had started, drawing over it.

"Hey, hey—stop it!" she said, taking the sketchbook back. But Junji felt embarrassed at having his face drawn, so he stole the book back from her and added some more to the sketch. Kiriko leaned forward this time, but she was evidently unable just to sit and watch—she couldn't just leave the drawing to him, it seemed—and she took the sketchbook back again, starting once more to draw. This same cycle of taking and having taken was repeated over and over, and in this manner Junji's face continued to be drawn. The outline of his face grew blurry in certain places, places where the lines Kiriko had drawn and the lines Junji had drawn overlapped excessively. Even a few unnecessary shadows had appeared. But all during the time they drew Junji's face together, a warm affection for Kiriko had been welling up inside of him. It seemed to him that even the drawing expressed this emotion. He had stopped feeling the shyness he had felt at first at having his face sketched—indeed, drawing over the drawing Kiriko had begun filled him with pleasure, as though they were laying the hands of their hearts one on top of the other.

"Well, it's finished." Kiriko stopped drawing and looked back and forth from the sketch to Junji's face, comparing the two. "It does look a bit like you, doesn't it?"

"Here, let me draw a little more."

"Where? Around the eyes?"

"It's my face—if I don't finish it myself . . . "

"You're awfully sure of yourself, aren't you."

"No. But—why did you draw my face?"

"Because we're coming from drawing practice, I would think. But also because when I started drawing I kept being reminded of my late daughter. She was just the right age to marry someone like you. I had her when I was nineteen—she was my only child."

". . . . ."

"Of course, I thought about her even when I looked at that model. Her body wasn't very pretty—I really didn't even want to draw her. But it was fun drawing you."

"You'll have to let me draw your face next time, assuming that we can go home on the same train after the next class."

Kiriko did not respond to this.

"If my daughter were alive she would have been able to meet you too." Grief hovered in Kiriko's eyes as she stared at Junji's face.

"She had never known love—she died just when the bud of her flower was starting to open. And I think that must have been best for her. . . . Maybe that's what happiness is?"

"I didn't think people had any way of knowing whether they're happy or unhappy once they die. Don't you think the people left behind just go ahead and think whatever they like—that they decide for themselves whether the person who died was happy or not?"

"You do have an unpleasantly logical way of thinking, don't you? You know, near the end of winter, when spring was just beginning, my daughter used to wake up in the morning and say—Ah, this is so much fun!—and then she'd stroke her arms. During the course of a single night her skin would turn silky smooth. That's the age she died at."

". . . . ."

Returning home on the day of the next drawing class, Kiriko suggested that they not go straight to Shinbashi station—she invited Junji to go with her to a department store. She bought him a ready-made suit; she seemed to think that they would stand out even more if Junji wore his school uniform.

The things Kiriko said to him didn't sound very affectionate, either—even when they were in the room where they went to be alone. "I'm sorry," she said. "It's just that you're the perfect age to marry my daughter." Still, in her pleasure Junji came to know the pleasure of being a man. It was an awakening that overflowed with strength. After a time, in a flirtatious voice that disguised her shame, Kiriko said, "I was thinking this before, when we were buying the clothes, but—you're tall, aren't you? Put your legs together for a second. . . ." She felt around for Junji's heels with her own, then pressed her face into his chest. "Look, I only come up to here."

She lay still, as though savoring the moment.

Kiriko didn't show up at the next week's class in Western drawing. Junji telephoned her house and asked to speak with her.

"Why didn't you come to class today?"

"The second we met everyone would know—the way you'd act would give us away. There's no way you'd be able to hide it."

They arranged to meet somewhere else for their third date, but Kiriko didn't show up at the appointed time. Junji called again.

By the time he took Kiriko's earlobe in his mouth, Junji had begun to feel both uneasy and irritated himself. Hadn't she just been dragged along by the fact of what they had done together that first time? Wasn't that the only reason she kept coming to meet him? And wasn't it Junji who dragged her along, forcibly? Did she have any choice in the matter? Even Junji could feel that her body was more tightly closed to him than it had been at first.

They had drawn Junji's face together, then they had lain with their heels aligned—had that been the last of the pleasure she felt with him? Had Kiriko felt nothing since then but an increasing pain, an ever growing sense of self-reproach?

Everything seemed to have happened almost as soon as they had met, and so at first Junji had given no thought at all to Kiriko's husband. But after a time he began to be jealous, and with this jealousy he acquired a sense of his own sinfulness.

"How old is your husband?" Junji asked. These were the first words he said that had anything to do with Kiriko's husband.

"Fifty-two. Why, does it matter?"

"I can't imagine you living with someone fifty-
two years old."

". . . . ."

"He commutes to Tokyo?"

"Yes, he commutes."

"I might even have met him on the train—maybe
in the station. I bet I'll meet him sometime," Junji
said.

Kiriko's chest tightened suddenly.

"Why? Do you want to meet him?"

"I don't know the first thing about you—about
your mind or about how you live. . . . I don't have
the slightest bit of influence over you. I went to see
your house, you know—secretly."

"What?"

"I mean—I think it's best if I get a look at him."

"No! You can't do that! Look—why don't we
just stop seeing one another." Kiriko's voice shook,
and she spoke quickly. "Have I really made you so
sick?"

"Sick? . . ."

"Yes. I knew that I'd been hurt, but I really didn't
think that you had—at least not so badly. I'm only
telling you this because you raised the subject, but
my relationship with my husband . . ." She hesitated.

"Your relationship is what?"

"It's not like it used to be. As you said before, my
mind and the way I live. . . . My husband doesn't
seem to have noticed at all, but I've changed. We
women are no good."

"What do you mean no good? What do you mean you've changed?"

Kiriko couldn't answer his questions. Junji continued to kiss her body everywhere, in various ways, but Kiriko was still holding back. Her restraint filled Junji with a frantic emptiness.

And once this emptiness set in, he had even less of a choice—he had to telephone Kiriko.

She would be coming by the next train on the Yokosuka line. Junji kept imagining her, drawing her inside his head—he kept thinking through the sequence of kisses he would give her, imagining the methods he would use—and he was startled to find that he seemed to have more fun doing this than he did when he was actually with her. He began to wonder if he might not really be sick, just as Kiriko had said. He began to be suspicious of himself.

That night too, Junji began with her ears. He had yet to find anything wrong with her body in places other than her ears. He was still moving from place to place over Kiriko's body when she muttered, "You don't have to do that, you know." Suddenly Junji was unable to move. But Kiriko relaxed. She felt as she had that first time, when they lay with their heels together. When Junji realized that Kiriko had spoken as she had because she pitied him, tears suddenly spilled from his eyes and would not stop. He thought—Is this what it means to break up? And yet Kiriko's cruel words also seemed to suggest that Junji had been doing things that her husband didn't.

Yumiura

His daughter, Tagi, came to tell him that
there was a woman at the door who said
she'd met him thirty years earlier in Yumiura, Kyushu. Kōzumi Shōzuke thought for
a moment and decided that he might as
well have the woman shown into the drawing room.

Unexpected callers came almost every
day to see Kōzumi Shōzuke, the novelist—
even now there were three guests in the

drawing room. The three guests had come separately, but they were all talking together. It was about two o'clock on an afternoon unusually warm for the beginning of December.

The fourth caller, the woman, knelt in the hall just outside the door she had opened, evidently embarrassed before the other guests.

"Please, come right in," Kōzumi said.

"Oh—it's truly, it's truly . . ." said the woman, in a voice that almost shook. "It's been such a long time! My name is Murano now, but when we met my name was Ta'i. Perhaps you remember?"

Kōzumi looked at the woman's face. She was a little past fifty, but she looked younger than her age—her pale cheeks were tinged with red. Her eyes were still large, despite her age. No doubt this was because she hadn't grown plump in middle age, as people often do.

"Just as I thought—you are the man I met. There's no doubt about it." The woman's eyes gleamed with pleasure as she stared at Kōzumi. There was an enthusiasm in her gaze that was lacking in Kōzumi's own as he looked back at her, trying to remember who she was. "You really haven't changed at all, have you? The line from your ear to your jaw, yes, and there—the area around your eyebrows—it's all just as it was. . . ." She pointed out each of his features, one by one, as if she were reciting a description of someone missing or wanted. Kōzumi felt embarrassed and also slightly worried at his own lack of memory.

The woman wore a black *haori* with her family crest embroidered on it in places, and with this an unostentatious kimono and obi. Her clothes were all well-worn, but not enough to suggest that her family had fallen on hard times. Her body was small, as was her face. She wore no rings on her short fingers.

"Thirty years ago you came to the town of Yumiura—or perhaps you've forgotten? You were kind enough to come by my room. It was the day of the Harbor Festival, toward evening. . . ."

"Hmm . . ."

Hearing that he had gone to the young woman's room—there was no doubt that she had been beautiful—Kōzumi tried once more to remember her. Thirty years ago Kōzumi had been twenty-four or twenty-five, and not yet married.

"You were with Kida Hiroshi and Akiyama Hisarō. The three of you had stopped at Nagasaki— you were traveling in Kyushu. We invited you to attend a celebration that was being held in honor of the founding of a small newspaper in Yumiura."

Kida Hiroshi and Akiyama Hisarō were both dead, but in life they had been novelists some ten years Kōzumi's senior—writers who had befriended and encouraged him from the time he was twenty-two or twenty-three. Thirty years ago they had been novelists of the first rank. It was true that the two of them had spent some time in Nagasaki around then—Kōzumi remembered their diaries of those travels and anecdotes they had told about them, diaries and anecdotes that were certainly known to the literary public.

Kōzumi wasn't sure that he had been invited to go along on that trip to Nagasaki—he had just been starting out in the world—but as he searched his memories of Kida and Akiyama, those two role models of his, men who had influenced him so much, their faces rose up again and again before him—he remembered the numerous favors they had done him—he was drawn into a fond and tender mood of recollection. The expression on his face must have changed, for the woman spoke.

"You've remembered, haven't you?" she said. Her voice had changed, too. "I had just had my hair cut very short and I told you how embarrassed I was— how I felt cold from my ears to the back of my neck. Autumn was just about over, so it must have. . . . That newspaper had just been established in town, and I had gathered my courage and had my hair cut short so I could be a reporter—every time your eyes moved to my neck I'd turn as though I had been stung, to hide it from you. Oh, I remember it all so well! You came back to my room with me, and right away I opened my ribbon box and showed you the ribbons in it. I think I must have wanted to give you evidence—to prove that my hair had been long, that I had tied it up with ribbons until two or three days before. You were surprised at how many there were, but as a matter of fact I had always liked ribbons, ever since I was small."

The three other visitors remained silent. They had all discussed with Kōzumi the things they had come to discuss, but since there were other guests they had been sitting at ease, talking of this and that, when the woman arrived. It was only proper that they should pass Kōzumi on to this next visitor, allow him to speak with her—but there was also something in her manner that compelled them to be silent. Indeed, all three of the visitors avoided looking at the woman's face, or at Kōzumi's, and though they were able to hear what was being said, they tried not to look as if they were listening directly.

"When the newspaper company's celebration was over we went down a street in town, straight toward the ocean. There was a sunset that seemed it might burst into flame at any moment. I'll never forget what you said then—that even the tiles on the roofs looked crimson, that even the back of my neck looked crimson. I replied that Yumiura was famous for its sunsets, and it was true—even now I can't forget them. That was the day we met—that day, with its beautiful sunset. The harbor was small, shaped like a bow—it looked like it had been carved out of the coast, there just under the mountains—and that's why it was named Yumiura, the bow-inlet. The colors of the sunset all collect there, in that scooped-out place. The high, rippled clouds in the sky at sunset that day seemed to be closer to the ground than the clouds one sees elsewhere, and the horizon out on the ocean seemed strangely close—it looked like the flocks of black birds that were migrating wouldn't have enough room to make it to the other side of the clouds. The colors of the sky didn't really seem to be reflected in the ocean, it seemed like that crimson had poured down only into the small ocean of the harbor and nowhere else. There was a festival boat decorated with flags on which people were beating drums and playing flutes, and there was a child on the boat— and you said that if you lit a match near that child's red kimono the whole ocean and the sky would burst instantly into flame, with a whoosh. Do you remember?"

"Yes, I think maybe I do. . . ."

"Since my husband and I married, my memory for things has gotten so bad it's pathetic. I guess there's no such thing as being so happy that you can decide not to forget. I know that people as happy and as busy as yourself don't have the time to sit around thinking of dull days from the past, and of course you really don't need to. . . . But Yumiura was the nicest town I've ever been in—in my whole life."

"Did you spend much time in Yumiura?" Kōzumi asked.

"Oh no—I got married and went to Numazu just six months after I met you there. Our older child has graduated from college and he's working now, and the younger, our daughter—she's old enough that we're hoping to find her a husband. I was born in Shizuoka, but because I didn't get along with my stepmother I was sent to live with relatives in Yumiura. I was eager to find some way to rebel, and so as soon as I arrived I went to work for the newspaper. I was called back and married off when my parents found out, so I was only in Yumiura for about seven months."

"And your husband is . . . "

"He's a priest at a Shinto shrine in Numazu."

This was not the sort of profession Kōzumi had been expecting, so he glanced up at the woman's face. The word is outdated and may end up giving the reader an unfavorable impression of the woman's hairstyle, but her hair was arranged in a pretty "Fuji-style"★ cut. Kōzumi's eyes were drawn to it.

"We used to be able to live fairly well considering that he's a priest, but after the war things got tougher and tougher day by day until now—my son and daughter still stand by me, but they find all sorts of ways to defy their father."

Kōzumi sensed the disharmony of the woman's family.

"The shrine at Numazu is so big it doesn't even bear comparison with the shrine at that festival in Yumiura—and of course the bigger they are the harder it is to manage them. We're having some problems just now because my husband decided to

---

★A once fashionable hairstyle in which the bangs are cut to resemble the slope of Mount Fuji. Tr.

sell ten cedar trees that grew behind the temple without consulting anyone. I've run away—come here to Tokyo."

". . . . ."

"Memories are something we should be grateful for, don't you think? No matter what circumstances people end up in they're still able to remember things from the past—I think it must be a blessing bestowed on us by the gods. There were lots of children at the shrine on the road going down through the town in Yumiura, so you suggested that we just keep going without stopping in—but even so we could see there were two or three flowers blooming on the small camellia over by the toilet—flowers with what they call 'double petals.' I still remember that camellia sometimes, even now—and I think what a wonderfully gentle person whoever planted it must have been."

It was clear that Kōzumi appeared as a character in one of the scenes in the woman's recollections of Yumiura. Images of that camellia and that bow-shaped harbor rose up in Kōzumi's mind as well now, seemingly called up by what the woman said. But it irritated him that he could not cross over into that country in the world of recollection where the woman lived. The two of them were as isolated from one another as the living and the dead of that country. Kōzumi's memory was weaker than that of most people his age. He sometimes talked at length with people whose faces he knew, yet could not remember their names—in fact it happened all the time. The unease he felt at such times was mixed with fear. And as he tried unsuccessfully to call up his own memories of the woman, his head began to ache.

"When I think about the person who planted that camellia, it seems to me that I ought to have made my room there in Yumiura a little nicer. But I hadn't and so you only came that once, and then thirty years passed without us ever meeting. Though to tell the truth I'd decorated my room a little even then, to make it look more like a young woman's."

Kōzumi could remember nothing at all of her room. Perhaps wrinkles formed on his forehead, perhaps his expression became slightly severe, for the woman's next words showed that she was preparing to leave.

"I must apologize for having come so suddenly, that was rude. . . . But I've wanted to see you for such a long time, and coming here has really been such a pleasure for me—nothing would have made me happier. I wonder—would you mind very much if I came again sometime?—if I could—there are some things I'd like to discuss."

"That would be fine."

The woman's tone of voice suggested that there was something she was not saying, something she hesitated to say in front of the other guests. And when Kōzumi walked out into the hall to see her off, the instant he slid the door shut behind him, the woman let her stiff body slacken. Kōzumi could hardly believe his eyes. This was the way a woman held her body when she was with a man she had slept with.

"Was that your daughter earlier?"

"Yes."

"I'm afraid I didn't see your wife. . . ."

Kōzumi walked out into the entryway ahead of the woman without answering. He addressed her back as she bent down to put on her sandals in the entryway.

"So I went all the way to your room, in a town called Yumiura."

"Yes." The woman looked back over her shoulder. "You asked me to marry you. In my room."

"What?"

"I was already engaged to my husband at the time—I told you that, and refused, but . . . "

It was as though Kōzumi's heart had been pierced with a pin. No matter how bad his memory had gotten, to think that he should entirely forget having proposed marriage to a young woman—to be almost unable to remember that young woman—he didn't even feel surprised, no—it struck him as grotesque. He had never been the sort of young man to propose marriage lightly.

"You were kind enough to understand the circumstances that made it necessary for me to refuse," the woman said, her large eyes filling with tears. Then, her short fingers trembling, she took a photograph from her purse.

"These are my children. My daughter is much taller than I was, but she looks very much like I did when I was young."

The young woman was small in the photograph, but her eyes sparkled brightly and she had a beautiful face. Kōzumi stared at the picture, attempting to make himself remember having met a young woman like this some thirty years earlier on a trip and having asked her to marry him.

"I'll bring my daughter sometime, if you don't mind—then if you like you can see how I used to be." It sounded as though there were tears mixed in with the woman's voice. "I've told my son and daughter everything, so they know all about you. They speak of you as though you were an old friend. I had really terrible morning sickness both times, sometimes I got a little crazy—but then as the morning sickness started getting better, around the time when the child started moving—it's odd but somehow I'd start wondering if the child might not be yours. Sometimes in the kitchen I'd sharpen a knife. . . . I've told my children about all that, too."

"You . . . don't ever do that."

Kōzumi was unable to continue.

At any rate, it appeared that the woman had suffered extremely because of Kōzumi. Even her family had. . . . Or perhaps on the other hand a life of extreme suffering had been made easier for her by virtue of her memories of Kōzumi. Even her family had been affected. . . .

But that past—her unexpected meeting with Kōzumi in the town called Yumiura—had evidently gone on living strongly inside the woman, while in Kōzumi, who had committed a sort of sin, it had been extinguished, utterly lost.

"Shall I leave the picture with you?" she asked, to which Kōzumi replied by shaking his head. "No, don't."

The small woman walked with short steps through the gate and then vanished beyond it.

Kōzumi took a detailed map of Japan and a book in which the names of all the cities, towns, and villages in the country were listed down from a bookshelf and brought it back into the sitting room. He had the three visitors search for him, but neither he nor any of them was able to find a town by the name of Yumiura anywhere in Kyushu.

"It's very strange," Kōzumi said, looking up. Then he closed his eyes and thought. "I don't remember ever having gone to Kyushu before the war. No, I'm sure I didn't go. That's right—I was sent on an airplane to a base the Special Attack Forces had in Shikaya during the battle in Okinawa—I went as a reporter for the navy—that was the first time I went to Kyushu. The next time was when I went to see Nagasaki just after the atom bomb was dropped. That was when I heard the stories about Mr. Kida and Mr. Akiyama—about their having come there thirty years earlier, from people in Nagasaki."

The three guests put forward a number of opin-
ions about the woman's fantasies or delusions,
laughing all the while. The conclusion was, of
course, that the woman was crazy. But Kōzumi
couldn't help thinking that he was crazy, too. He had
sat listening to the woman's story, half believing and
half doubting that what she said was true, searching
his memory. It happened that in this case there was-
n't even a town called Yumiura, but who could say
how much of Kōzumi's past others remembered,
though Kōzumi had forgotten it himself—though it
no longer existed within him. The woman who had
come that day would almost certainly go on believing even after Kōzumi died that he had proposed to
her in Yumiura. Either way it was the same.

The Boat-Women: A Dance-Drama

I. KURETAKE'S HOUSE

KURETAKE	*A dancer*
MURASAKI	*Kuretake's daughter (ten years old)*
KAGEKIYO	*A man of the Heike clan*
FIVE DANCERS	
KABŪ	*A boy spy (twelve or thirteen years old)*
KOSASA	*Kuretake's servant, an old woman*

ONE OF KAGEKIYO'S ATTENDANTS
ONE OF KAGEKIYO'S RETAINERS

Kuretake's house. Downtown in the capital. The cherry trees in the garden are in full bloom. Dusk approaches. The curtain rises as the chorus sings the following verse.

〴 Buddha is ever present   but does not ever really appear   —how sad.

> (*The curtain rises.*)

Kuretake is teaching the steps of a dance to the five young dancers. They blossom like the garden's cherry trees, brilliantly.

〴 In the darkness before dawn   no human noise perhaps you can see him dimly in a dream.

FIRST DANCER: Even in a dream in the darkness before dawn—the figure of Buddha. . . .

SECOND DANCER: I've never been able to see him.

THIRD DANCER: The only thing I see even in my dreams is *his* face. . . .

They laugh.

FOURTH DANCER: And yet, they say there is a path to buddhahood. . . .

FIFTH DANCER: Even in the playful games of children.

KURETAKE: The lighthearted games of young children are precious indeed. I myself am spattered with the dark grime of this world—and yet when I awake in the middle of the night, dreaming in the dark, suddenly the motions of a dance, the melody of a song drifts up in my mind. And this—this is the same as being lit dimly by Buddha's light. . . .

FIRST DANCER: But Kuretake, you're famous as a dancer—known in the capital—people say you're superior even to Gojō-no-Otsumae. . . .

SECOND DANCER: Surely a golden Bodhisattva appears in your dreams, and the two of you sing together, and dance.

KURETAKE: Don't be foolish. . . . Otsumae was one of the greatest dancers of all time. She was summoned to the Imperial Palace after she turned seventy—she passed away in the spring of her eighty-fourth year, contented, listening to a poem intoned by the Emperor, who had come to visit her on her deathbed—how could you compare someone like me to her?

FOURTH DANCER: My goodness! (*Surprised.*) I wonder if we'll still be singing at eighty-four.

SECOND DANCER: Hotokegozen, though dearly loved by Lord Kiyomori, visited Giō in Sagano and became a nun at sixteen. . . .

THIRD DANCER: Giō's place in Lord Kiyomori's heart had been stolen by Hotokegozen, so at twenty she became a nun. . . .

FIRST DANCER: At nineteen her younger sister, Gijo. . . .

KURETAKE: In a hopeless world, wherein lies a woman's happiness?

ᄭ Flowers of many kinds blossom   fruit ripens
—how sad.

Urged on by Kuretake, the dancers dance.

〽 Wanting to play I was born   wanting to
frolic I was born   when I listen to the voices
of children playing   I am reminded of my
own spring   will it not scatter   the blos-
soming flower   reaching to take it   come
let's play.

> Kuretake's daughter, Murasaki, enters the garden as they
> dance. She frolics with (the young) Kabū.
> Kosasa sees them and gestures to Murasaki to come in.
> Kabū looks into the house.

KURETAKE: (*Turning to face Kabū, speaking firmly.*) Slander-
ers of the Heike Clan are not permitted to enter
this house. We have no use for spies.

KABŪ: It's dangerous with all the commotion in
town, so I escorted Murasaki home.

KURETAKE: What commotion? . . .

> Murasaki takes Kabū's hand, invites him to enter the
> house.

KABŪ: Heike warriors attacked the parade of the
Emperor's Chancellor as it was making its way to
the palace. Their violence was extreme. . . .

> The dancers are surprised. Kabū continues, speaking as
> though what he is saying is perfectly ordinary.

Of course, earlier—Lord Shigemori's son met the
Chancellor's carriage on his way home from his
flute lessons, and when he didn't greet him he was
whipped. It was terrible. But today, with this
revenge—I'm sure the nobles realize only too well
what the Chancellor did.

> Murasaki dances innocently by herself.

⌇ Standing gathering seaweed on the rocky
shore of Koyorogi   do not wet those young
women   waves stay offshore   waves stay
offshore.

Kabū frolics, seems to be tangled up in Murasaki's dance.
Kagekiyo comes into the garden.

KOSASA: Kagekiyo has arrived.

Kabū sees Kagekiyo and flees.
The dancers stand and begin to leave.
Kagekiyo prevents them from leaving, then sits down.

KAGEKIYO: I won't let you go home until you sing a
verse.

DANCERS: Yes.

⌇ I yearn to see you   my love I yearn to see
you   I'd be thrilled if we could meet   if
only we could meet   if I could see you   if
only we could meet.

The dancers dance. Murasaki dances playfully in their
midst.

KAGEKIYO: Your Murasaki's dancing won't embarrass
you . . . she's become beautiful.

KURETAKE: Well, who can say. . . . She's past ten
already—it seems there are some boys who tease
her. It's quite pitiful, as it says in the poem.

KAGEKIYO: What? In this world, with the Heike so
rich and powerful—my own bravery more
resplendent than ever. . . . Well, Kuretake, shall we
have a dance?

KURETAKE: Very well.

Kagekiyo takes up a biwa and begins to play.

Kuretake dances in time to Kagekiyo's biwa. Murasaki
watches, absorbed.

〜 Oh how precious　　the preciousness of this
day　　this life the life of a drop of dew yet still
I chance to meet　　the joy of this day.

Kagekiyo stands and dances in time with Kuretake.

〜 The sadness of this day　　yesterday a dream
tomorrow an illusion　　today in reality here on
my lap　　the biwa I pluck and make sing
whose child listens　　it is my own good child
oh how precious　　the preciousness of this day.

An evening breeze blows up as they sing and dance. The
air fills with falling petals. The sun begins to sink.
Kagekiyo's retainer enters the garden.

RETAINER: Sir—

Coming out into the garden, Kosasa, the old woman,
prevents him from speaking.

RETAINER: (*To the old woman.*) Tell his lordship that the
troops have all been ordered to gather this
evening—we're off to destroy the Genji of the
eastern lands. . . . I came as fast as I could. . . .

The old woman is surprised. She has the retainer leave.
Looking at Kuretake, who is still dancing with Kagekiyo,
the old woman has a premonition. It seems to her that
something terrible will happen.
Murasaki becomes involved in the dance, dances inno-
cently with Kagekiyo.
Kagekiyo too feels vaguely uneasy. He dances all the
more intensely with Kuretake.
The old woman gazes at the dancers, absorbed in
thought.

(*Curtain.*)

2. KURETAKE'S RUSTIC RESIDENCE

KURETAKE

MURASAKI   *About fifteen years old*

TOKIMARU   *Formerly Kabū, now about seventeen*
                years old

OLD WOMAN KOSASA

Kuretake's grass hut in the bamboo forest at Saga.
A summer night. The moon shines brightly, quietly.

〰 Even if it is not so   in the dew-wet fields of
Saga. . . .

(The curtain rises.)

Murasaki stands by herself.
Kosasa sets out to fetch water from the well.

〰 Longing for the past in a grass hut   still
faithful this bamboo   a single drop of dew
on a bamboo leaf   holds the light of the
moon   but this life too will fall   —how sad.

The sound of a flute approaches through the bamboo
forest.
Tokimaru (formerly Kabū) discovers Murasaki as he
makes his way through the bamboo and runs out into
the open.

TOKIMARU: Murasaki *(he calls)*—it's more difficult for
a person hiding from the world to search for a
person hiding from the world than it was for
Kabū to search for people who slandered the
Heike.

MURASAKI: *(Walking towards him, nostalgically.)* Tokimaru.

TOKIMARU: Murasaki—I only wanted to meet with you once. The fighting in town frightened me, so I walked through a village in the mountains, playing my flute, yearning for you.

MURASAKI: My heart was charmed by the beautiful sound of your flute.

TOKIMARU: The sound of the flute led me on to this place, where a person I knew long ago resides. I swore that if I met you I would stop playing. Here, let me give you this flute as a memento. I have nothing to regret now—now that I have met you once.

MURASAKI: Why do you keep saying "once"? . . .

TOKIMARU: The haughty Heike, too, lasted but a moment. . . .

⋏ The mighty must fall   they who are together must part   a single brief meeting is the same as a bond of fifty years. . . .

Tokimaru gives the flute to Murasaki and holds her hands, reluctant to part with her.

TOKIMARU: I'll climb Mount Hiei tomorrow. I prayed as I played my flute that the day I met you would be the day I became a priest.

The old woman returns, having filled a bucket with water. She sees them.
Murasaki and Tokimaru draw apart.

OLD WOMAN: Murasaki, you mustn't leave your mother's side even for a moment.

She goes into the house and arranges gentians, miscanthus, and other plants in the bucket.

TOKIMARU: How is Kuretake?

MURASAKI: In this life there's no way of knowing what tomorrow will bring. . . . She's worn out from thinking of Kagekiyo. . . .

TOKIMARU: Is she ill?

MURASAKI: She can't bear to play the biwa, even, linked as it is to the Heike—she can no longer dance or sing—her life has collapsed completely. She's shameful to look at.

TOKIMARU: No, I'm the shameful one. Horses and carts moved off to the side when they passed me, with my hair cut short, dressed in red *hitatare*—and I thought it was because the way I carried myself gave me an air of authority. I walked proudly about town, burst into ordinary people's houses, captured people. To undo that sin. . . .

OLD WOMAN: Murasaki, Murasaki.

> She calls Murasaki into the house. Murasaki goes in sadly, reluctant to part with Tokimaru.

TOKIMARU: Murasaki. . . . I'll be a priest after tomorrow—your face will be my image of the Bodhisattva. I'll worship it day and night.

> Tokimaru begins to head back in the direction from which he came, but Murasaki comes out of the house and goes after him.

⌃ In this ephemeral world   just once I was able to meet   a person dear to me. . . .

> Tokimaru walks off into the bamboo forest. Murasaki follows after him for a while, then returns with the flute tucked into her obi.

OLD WOMAN: Murasaki, Murasaki.

> Murasaki returns to the house.
> Kosasa removes the screen. She lights the lamps.
> Kuretake is lying in her bedding, sick.

〽 I yearn to see you   my love I yearn to see
you   I'd be thrilled if we could meet.

KURETAKE: Murasaki! We must be grateful—the
light of the Buddha came streaming in. For me,
a dancer, the light of the Buddha is the light of
art. . . . As long as I keep my eyes focused on
that single ray, I have no troubles—I'm not sick.
I begin to be able to hear the sound of
Kagekiyo's biwa, strains that rise from those four
strings.

> Kuretake rises and takes up the biwa. It reminds her of
> the past. She embraces it.

〽 You drift on the waves of the western sea
afloat in a boat   when a wind blows. . . .

KURETAKE: I don't know where Kagekiyo has gone.

〽 After he had gone   the rustle of leaves in the
clump of bamboo   a lonely orphan's ties
should be   with a distant father.

> They all cry, heads sunk on their chests.
> Kuretake sits up straight.

KURETAKE: When your mother is gone you must
search for the sound of your father's music.
MURASAKI: Yes, mother.
KURETAKE: You must dance this dance, as I teach
you.

Kuretake takes her fan and stands up straight. She tries with all her heart to remain standing, though her legs are unsteady.

OLD WOMAN: Go on, Murasaki . . . do your best to remember the dance your mother is about to show you—go on, Murasaki.

Murasaki takes her fan and stands ready, concentrating intensely.

〽 Oh how precious   the preciousness of this
    day   this life the life of a drop of dew yet still
    I chance to meet   the joy of this day   the
    sadness of this day.

Murasaki collapses. The old woman runs to her and lifts her up.

KURETAKE: I'll be watching you from the Pure Land when you dance to the music of your father's biwa, Murasaki.

Murasaki is possessed by her mother's spirit. She dances more and more beautifully, more and more desperately.

〽 Yesterday a dream   tomorrow an illusion
    today in reality   here on my lap   the biwa
    I pluck and make sing   someone's child lis-
    tens   it is my own good child.

Kuretake grows progressively weaker, then stops breathing.
The old woman clings to her, weeping.
Murasaki continues to dance all the more desperately.

(*Curtain.*)

3. AKI–NO–MIYAJIMA

KAGEKIYO	*Now blind (disguised as a* biwa-hōshi*)*
MURASAKI	*Dressed in travel costume*
THREE GENJI SAMURAI	
A PRIESTESS OF ITSUKUSHIMA SHRINE	
CROWDS OF MEN AT THE FESTIVAL	
CROWDS OF WOMEN AT THE FESTIVAL	

A place along the open corridor that circles Itsukushima Shrine. It is the day of the Autumn Festival, and the autumn leaves are beautiful. The crowds of men and women who have come for the festival are dancing. Bugaku music continues for a short time, then the chorus sings this verse.

〽 Along a path which drifts on waves across the sea the sighted and the blind in a single boat adrift in a single fate.

(The curtain rises.)

〽 Yesterday we flourished with flowers in the capital today we've come to the autumnal western seas Aki-no-Miyajima the solemn Heike palace though he is blind Kagekiyo is manly and brave excellently elegant the notes he plays the familiar tune tears fall on this biwa of mine harari hararin harari hararin the lap on which the biwa rests has grown old.

Men and women from the festival pass by.

The blind Kagekiyo leans alone on the railing of the corridor, his biwa on his back. He crouches down and listens to the music without moving.
The music stops and from the direction from which it had come three Genji samurai appear, walking down the corridor.

FIRST SAMURAI: The beautiful priestesses on this beautiful day. . . .

SECOND SAMURAI: Long ago a priestess from Itsukushima was summoned to the capital, and at Lord Kiyomori's mansion. . . .

THIRD SAMURAI: She was so lovely she was permitted to dance even in the presence of the Cloistered Emperor Goshirakawa.

SECOND SAMURAI: The bugaku today was dedicated to the gods by Lord Yoritomo. . . .

THIRD SAMURAI: He's going off to fight Yoshitsune in Oshō, he prays that he'll win the battle. . . .

They draw nearer to Kagekiyo, and speak suspiciously.

FIRST SAMURAI: Well, well—a *biwa-hōshi*.

SECOND SAMURAI: Why would a *biwa-hōshi* be off by himself, sunk deep in thought? Vulgar, isn't it—that ferocious expression. . . .

THIRD SAMURAI: A suspicious man . . . is he a Genji or a Heike?

KAGEKIYO: I cannot see the Genji world. . . .

THIRD SAMURAI: What?

KAGEKIYO: You'll notice that I'm blind.

SECOND SAMURAI: What can't you see—say that once more?

KAGEKIYO: I do not listen to the thing they call the Genji biwa.

During this exchange some of the people from the festival come across the corridor from the other side, appearing one at a time, and surround Kagekiyo and the samurai.

FIRST MAN: Why, it's a Genji interrogation.

PEOPLE: An interrogation, an interrogation!

FIRST WOMAN: They arrest a blind *biwa-hōshi.*. . .

SECOND MAN: He said he hasn't got a Genji biwa.

Everyone laughs.
First Samurai looks over the women in the crowd.

FIRST SAMURAI: Are there any Heike women? A woman's better than a blind man.

The women flee.

THIRD SAMURAI: At a Genji festival anything with a "Heike" in its name pollutes the shrine—get rid of that biwa.

He pushes Kagekiyo away from the railing.

THIRD MAN: Is there even such a thing as a Genji or a Heike at a festival?

FOURTH MAN: It's the people's festival, they're the people's songs.

PEOPLE: Dance, dance!

SECOND SAMURAI: What?

The samurai look angry, the people of the festival jeer.
The people dance, as though mocking the samurai.

〳 Genji and Heike   aristocrats and warriors
    parents and children and brothers and brothers
    attack one another   war is the people's
    agony   women and children flee   today
    the town burns   and the looting. . . .

As the chorus sings this verse the third samurai takes
hold of the front of Kagekiyo's cloak. Kagekiyo automatically assumes a fighting position and twists the samurai's
arm. The samurai show signs of fear.
The people laugh.
The three samurai become angry and drag Kagekiyo to
the center of the stage.
The people jeer with increasing boisterousness.

FIRST MAN: May you turn into a three-horned devil!

FIRST SAMURAI: What!

SECOND SAMURAI: You dare defy us?

The samurai chase the people, who scatter. The people
laugh, clown about, and dance as they scatter.
The samurai return and address Kagekiyo.

THIRD SAMURAI: If you're really a *biwa-hōshi*, sing for
us.

SECOND SAMURAI: Yeah, hurry up. . . .

FIRST SAMURAI: Sing.

Kagekiyo has no choice, so he plays.

〽 Here on my lap   the biwa my hand knows
    so well   I play it   I make it sing   ah
    what a cheery song   whose child listens   it
    is my daughter.

The three samurai and the people from the festival all
listen.
Kagekiyo stops playing and stands.

KAGEKIYO: The biwa has many modes, the heart has
many registers, harari hararin.

The samurai mimic Kagekiyo.

FIRST SAMURAI: It is my daughter, harari hararin.

The samurai depart, laughing.

Kagekiyo goes off in the opposite direction.
The people watch him leave.

⌄ The morning glory   eaten by insects   the
fate of flowers   this too when autumn comes
the sighing of distant fields   the bank of a
river   in my dream as I sleep   I yearn to
meet my father.

Murasaki comes dressed in travel costume from the
direction where the music had been earlier, with the
flute tucked into her obi. She walks down the corridor
with the priestess of the shrine, looking sorrowful.

MURASAKI: I won't mind even if I have to travel a
thousand miles to meet my father, the man I
search for. But I thought if I became a priestess
here at Miyajima I would certainly meet him—
that's why I came.

PRIESTESS: (*Apparently unable to comfort her.*) How very
sad. It sounds as though you're from the capital—
and you're so beautiful. Even if you hope to
dance, it won't be allowed.

MURASAKI: You were kind enough to show me that
elegant dance—even that was a comfort. This
journey is so terrible, your dancing was a positive
joy.

Murasaki fusses with her clothes, as though unwilling to
part with the priestess.

⌄ Autumn drizzle on a mountain path   snow
on a road near the coast   even these breasts
of mine freeze   as I travel alone. . . .

The samurai return. Seeing Murasaki, they draw near
her.

FIRST SAMURAI: Wow, beautiful. I haven't seen you around here, young lady—are you a Genji firefly?

SECOND SAMURAI: Or a Heike bell cricket?

PRIESTESS: (*Thrown into confusion, shielding Murasaki.*) She has asked to be made a priestess, she's a guest. . . .

FIRST SAMURAI: No, she's not the kind of girl to dance before a god. She looks like a dancer, like she plays with men. Play your flute for us.

THIRD SAMURAI: Flute? . . . First that biwa, now this flute—these travelers with their musical instruments seem a bit strange. And she's too beautiful—it's suspicious.

He moves to catch hold of Murasaki.

FIRST SAMURAI: Hey, hey. Play your flute for us. Sing for us. If you dance well we'll let you be a priestess at the shrine.

MURASAKI: Will you really?

FIRST SAMURAI: Go on, hurry up and dance.

Murasaki dances.

⌁ Still I chance to meet the joy of this day
the sadness of this day yesterday a dream
tomorrow an illusion today in reality here
on my lap.

People from the festival appear one by one as she sings. They watch.
The samurai mimic Murasaki as she dances.

FIRST SAMURAI: Yeah, you said it—in reality here on my lap.

MURASAKI: Will you let me be a priestess here?

SECOND WOMAN: That song on the biwa, earlier. . . .

THIRD WOMAN: This one was just like it. . . .

MURASAKI: What? What do you mean, a biwa song just like it. . . .

FOURTH WOMAN: A song a traveler sang earlier.

MURASAKI: That traveler. . . . What kind of person was he? Which way did he go?

WOMAN 1, 5, 6: He went that way just a moment ago.

Murasaki heads hurriedly in that direction.

⌇ Looking out over the mountain scenery more beautiful than I had heard   Itsukushi- ma.

Murasaki hurries on as this verse is sung, until a samurai calls out to her.

FIRST SAMURAI: It was this way—if you're looking for the traveler, go this way.

He points in the opposite direction, winking at samurai 2 and 3.

SAMURAI 2, 3: Right, this way, this way.

MURASAKI: Was it really?

Murasaki starts to go in the opposite direction.

PEOPLE: That way, that way.

Murasaki heads back again.

SECOND SAMURAI: Hey!

SAMURAI 1, 3: This way, we're telling you this way.

The samurai lead Murasaki forcibly in the direction opposite from the one in which Kagekiyo went.

PEOPLE: But—

The people start to follow, but the first samurai glowers at them.

SAMURAI 1, 2: Dance!

PEOPLE: Dance!

The people from the festival dance peacefully.

〵 So very numerous   the votive tablets   the sound of bells on a priestess dancing a bugaku dance   in the rising tide lanterns   not autumn stars reflect   in the autumn-dyed mountains   a deer cries   I bow my head in prayer   with a quiet heart   I'll go from island to island   I board the boat   from the sandbank at the torii   untie the line that strings us to the shore   and push off   the oars' sweep beats a single rhythm   a peaceful cheery song.

The dancing people move away as they dance.

Then, as if nothing has happened, Kagekiyo appears on the corridor. He walks from stage right to stage left. When he reaches the middle of the stage the lights are extinguished.

(*Curtain.*)

4. TOMO INLET

KAGEKIYO

MURASAKI

TWO BOAT-WOMEN

THREE TRAVELERS

TWO BOYS FROM THE VILLAGE

ONE GIRL FROM THE VILLAGE

An inland sea, the harbor at Tomo. The ocean off in the
distance, a strait with no mouth. A small island closer to
shore. The sky is overcast, it looks as if it might begin to
snow at any minute. Shikoku is not visible.
The boat-women's small boat is near the shore.
A light snow starts to fall. Children from the village
dance playfully.

ᕋ In the inlet at Tomo   women divers fish for
bream   in the sea at Tomo   women divers
draw in their nets   they are so dear   they
are so dear.

ᕋ For their sisters   women divers fish for
bream   for their sisters   women divers
draw in their nets   they are so dear   they
are so dear.

(*The curtain rises.*)

ᕋ The fish climb   the wind blows   striking
the bucket-drum   striking the bucket-drum
how I wish it would clear   how I wish it
would clear.

Kagekiyo walks on stage while the children are dancing,
then stands still.

GIRL: Look! A *biwa-hōshi*, a *biwa-hōshi!*
FIRST BOY: Gosh, it's sad, isn't it—a blind priest.
KAGEKIYO: What's sad? The eyes of my heart see the
    things I want to see precisely as I want to see
    them. The flourishing capital, bravery at war, my
    own vigorous form. . . .
SECOND BOY: Tell us a story about that brave
    war. . . .

KAGEKIYO: And I can hear the sound of music from heaven.

GIRL: Play your biwa for us.

FIRST BOY: The one that goes beron-beron is the story of the Heike clan. . . .

GIRL/SECOND BOY: Barari-karari, karari-barari. . . .

The children help Kagekiyo take the biwa from his back. Kagekiyo sits in a formal posture and begins to chant.

KAGEKIYO: I myself am a general of the Heike clan, known by the name Aku Nanahei. . . .

That name roars   too valiant beyond comparison   Kagekiyo. . . .

KAGEKIYO: The Heike's luck ran out at war, it is painful even to remember—in the fourth year of Juei at Dan-no-Ura huge numbers of Genji troops overwhelm the Heike boats. . . .

A nun of the second rank   the Emperor's grandmother   always ready   she holds his majesty in her arms   and steps to the side of the boat.

KAGEKIYO: The Emperor is only eight years old, he asks—Nun where are you taking me?—You were born the leader of ten thousand carriages of war, but now your good fortune has ended. First face the east. . . .

Bid farewell to the great shrine at Ise. . . .

KAGEKIYO: Then face the west and pray to be carried off to the Western Pure Land, then pray to

Amitabha Buddha. The Emperor puts his dear little hands together. . . . (*Kagekiyo sobs.*) Now at last the nun holds him in her arms, saying—Beneath the waves there is another capital. . . .

⋏ Preparing to enter the sea   a thousand fathoms deep   my eyes   how can they see this.

KAGEKIYO: I cry out and crush my own eyes, my own eyes I. . . .

⋏ Crush.

The children are silent and still. They are looking at Kagekiyo's eyes. Kagekiyo forces himself to be calm.

KAGEKIYO: Well then, the story of a brave war. . . . At the Yumi river in Yashima, the reverse oar, the fan target.

The children look bored. Once more they frolic and dance.

⋏ Snow streams down   hail streams down.

CHILDREN: The boat's come, the boat's come!

The children scatter in the direction of the boat, which is not visible.

⋏ It falls and   it falls and   still it falls   how I wish it would clear   how I wish it would clear.

Kagekiyo walks off, holding the biwa to his chest. Two boat-women pass him. It is clear from their expressions that they are waiting for customers.

⋏ The fish climb   the wind blows   striking the bucket-drum   striking the bucket-drum.

Two travelers come from the boat and tease the boat-women.

SECOND CUSTOMER: I've heard that if one passes by Tomo offshore the cypress-wood fans beckon. . . .

FIRST BOAT-WOMAN: Tomo's most famous product—boat-women. . . .

FIRST CUSTOMER: Those cypress-wood fans. . . .

FIRST BOAT-WOMAN: Adorn the cabins on the boats. . . .

FIRST CUSTOMER: The scarlet collars. . . .

SECOND BOAT-WOMAN: Are wrapped around our waists. . . . (*She lifts the hem of her kimono and displays her legs.*)

SECOND CUSTOMER: Boat-women are. . . .

⌁ Pillowed on the waves   rocking in a small boat   lovely and swaying.

FIRST CUSTOMER: Out on the sea, in this snow?

⌁ Snow does not grow deeper on the sea   only the thoughts of women grow deeper.

FIRST CUSTOMER: Perhaps to see the cypress-wood fans. . . .

SECOND BOAT-WOMAN: Let's go to the boat—it's that one, that boat there.

SECOND CUSTOMER: Oh, it's cold. (*He looks at the boat and shivers.*)

FIRST BOAT-WOMAN: Shall we warm you with our snow-white skin?

SECOND CUSTOMER: Snow-white skin, did you say? It looks like goose-flesh to me, like the skin of a

shark come up from the sea. I was amazed when
I heard that Heike women had become prosti-
tutes—what a story. . . .

FIRST CUSTOMER: This ugly-faced Heike crab, this
spider-prostitute. . . .

SECOND BOAT-WOMAN: Weren't the Heike pinched
the way a Heike crab pinches?

FIRST BOAT-WOMAN: Weren't the Heike embraced
the way a spider-prostitute embraces?

CUSTOMER 1, 2: Oh! How frightening!

〽 The fish climb　　the wind blows　　striking
the bucket-drum　　striking the bucket-drum
how I wish it would clear　　how I wish it
would clear.

The second customer clowns about with the second
boat-woman, and the two of them board the boat. The
second boat-woman seems filled with power as she rows
out. The first customer draws closer to the first boat-
woman, and they walk off stage right.
Murasaki's clothing is disheveled, but she has not lost her
elegant and refined air. She chases the third man onto
the stage.

〽 In the inlet at Tomo　　women divers fish for
bream　　in the sea at Tomo　　women divers
draw in their nets　　they are so dear　　they
are so dear.

THIRD MAN: I told you my wife and kids are waiting, and you. . . .

MURASAKI: (*Taking the man's hand in hers.*) Even my
breasts are cold in this snow. . . .

THIRD MAN: Which one? (*He puts his hands on Murasaki's
breasts.*)

MURASAKI: Even these breasts sleep in pairs, on this
lonely chest of mine. . . . (*She leans coquettishly on him.*)
THIRD MAN: There's nothing odd in that—you've
got two.

The man shakes Murasaki off and starts to go. Murasaki
runs after him.

⌇ For their sisters   women divers fish for
bream   for their sisters   women divers
draw in their nets   they are so dear   they
are so dear.

MURASAKI: I'm not missing anything. When we're
doing it. . . .
THIRD MAN: You certainly aren't missing anything—
you're beautiful. (*He clasps her to him suddenly, without
thinking, then hesitates.*) You're a Heike prostitute
aren't you?
MURASAKI: Yes. After the Heike clan was defeated at
Yashima, when everyone was escaping—I was left
behind at the harbor in Tomo. . . .
THIRD MAN: How old were you then?
MURASAKI: Why—I must have been. . . . (*She thinks.*)
THIRD MAN: You can't say, can you. That mouthless
strait. . . .
MURASAKI: Won't you stay? Have a little rest in a
strait with no mouth, be left with no regrets.
THIRD MAN: Your mouth and your lies have both
gotten good. Yes, it seems the women are all
Heike throw-aways.
MURASAKI: You think the baby bird is pretty, and it
wants to be held by the daddy bird. . . .

The third man gives Murasaki some money. Murasaki makes a face showing that she thinks the amount too little. The man starts to go. Murasaki runs after him, surprised.

THIRD MAN: Look, I'm giving it to you.

MURASAKI: No, no—I don't need people's charity. . . .

THIRD MAN: Let me go. . . .

MURASAKI: I may have fallen in the world, but I don't beg.

THIRD MAN: You *are* a beggar, and you're stubborn—it doesn't go well with your face. Watch it—I'll throw you into the sea!

The man pushes Murasaki down.

MURASAKI: Well, of all the rough. . . . I don't beg. I may sell my body, but I don't beg.

Murasaki tries once more to stop the man. He shakes her off and pushes her down.

THIRD MAN: Grrh!

Kagekiyo pushes Murasaki and the third man apart and stands between them.

KAGEKIYO: (*To the third man.*) Don't be violent.

THIRD MAN: Yeah, which one of us being violent? My family's seen that the boat has come in.

The man hurries off.
Kagekiyo feels for Murasaki with his hands, shields her. . . .

KAGEKIYO: Are you hurt? Have your sleeves come unsewn? Oh, your hands are cold.

MURASAKI: (*Shaking him off, repulsed.*) A blind man will sew up the torn stitches for me?

KAGEKIYO: With a needle of the heart, tears in the heart. I'll sew them.

MURASAKI: The road we walk in this world is a mountain of needles—be very careful where you step.

Murasaki picks up the money the man gave her and gives it to Kagekiyo.

There was a beggar here earlier.

Murasaki leaves, making a show of her slovenliness.

⟍ In the inlet at Tomo   women divers fish for bream   in the sea at Tomo   women divers draw in their nets   they are so dear   they are so dear.

KAGEKIYO: How strange—that woman seemed. . . . She seemed to be a Heike, one of those left behind. (*He sits.*) The beautiful Heike have died out—I'll play for the wanderers.

He plays his biwa.

⟍ Oh how precious   the preciousness of this day.

Murasaki returns, seemingly possessed by the sound of the biwa. Then, feeling as though her heart has been pierced, slowly remembering the last dance her mother taught her, she begins to dance.

⟍ This life the life of a drop of dew   yet still I chance to meet   the joy of this day.

At last Kagekiyo stands and begins to dance. They fall naturally into line.

〽 The sadness of this day   yesterday a dream
tomorrow an illusion   today in reality here
on my lap   the biwa I pluck and make sing
whose child listens   it is my own good child
oh how precious   the preciousness of this
day.

Kagekiyo and Murasaki realize that they are parent and child. They want to tell one another their names, but they are unable to—they go on dancing. Snow falls so heavily that the two figures can no longer be seen.

〽 I try to tell you who I am   but I am filled
with shame   shall I say shall I not say   oh
father   oh daughter   and so together they
dance   together they dance   a chance
meeting   is itself   the fruit of an eternal
bond.

Kagekiyo moves as if to shake himself free of Murasaki, then clasps the biwa to his chest and cries. Murasaki places her own long cloak over Kagekiyo's shoulders and then continues dancing, seemingly even more possessed than before. She continues to stand in the fiercely falling snow.

(*Curtain.*)

Printed in the United States
by Baker & Taylor Publisher Services